THE CAPE

Military Space Operations

1971-1992

by Mark C. Cleary

45th Space Wing
History Office

Table of Contents

Preface

This is the companion volume to an earlier work, "The 6555th: Missile and Space Launches Through 1970." It presents the Air Force's space launch operations at Cape Canaveral Air Force Station (CCAFS) from the beginning of 1971 through 2 July 1992. Though the focus of this monograph is the 6555th Aerospace Test Group's operations at the Cape, the work places those activities within the larger context of U.S. Air Force space operations conducted under two major commands: Air Force Systems Command and Air Force Space Command. The last chapter presents a brief review of NASA, Defense Department and Eastern Space and Missile Center (ESMC) studies that address future space operations at the Cape.

This is an unclassified monograph, so details concerning launch vehicle performance on classified space flights are limited by strict security guidelines. References to other aspects of classified space programs are presented in their officially releasable context. Many of the 6555th's efforts involved unclassified programs (e.g., DSCS, NAVSTAR II, FLTSATCOM, NATO III, etc.), so the details of those operations can be presented as openly as possible. Thus, while security considerations have limited our story somewhat, we believe the reader will come away with an objective picture of the Cape's military space operations after 1970.

This monograph could never have been written without the conscientious efforts of Mr. Marven R. Whipple, the Air Force Eastern Test Range history staff, and the dozens of officers, airmen and civilians who contributed historical reports on the 6555th's various offices, branches and divisions. Five other people deserve special thanks for their efforts: Ms. Jan Crespino printed the initial drafts of the monograph; Lt. Colonel Richard W. Sirmons (USAF Reserve) did the formatting and layout of the work and printed the camera-ready proofs; Mr. Jeffrey Geiger provided key photos from Vandenberg AFB; Dr. Timothy C. Hanley and Dr. Harry N. Waldron, III reviewed the manuscript and provided additional photographic support. Dr. Hanley and Dr. Waldron also deserve special recognition for their comprehensive treatment of space operations in nearly two decades of SAMSO, Space Division, and Space Systems Division histories. All the aforementioned individuals have provided valuable services to the Air Force History Program and the heritage of the Air Force at large.

MARK C. CLEARY

January 1994

USAF Space Organizations and Programs

Air Force Systems Command and Subordinate Space Agencies at Cape Canaveral

Any examination of military space operations at Cape Canaveral should include at least a brief review of the original initiatives, institutions and programs that promoted those efforts from the 1940s onward. In that regard, the Air Force's early interest in space operations was sparked by discussions with the Navy shortly after the end of World War II. At Major General Curtis E. Lemay's request, the Douglas Aircraft Company's RAND group provided the Pentagon with a 321-page study in May 1946 on the feasibility of satellites for military reconnaissance, weather surveillance, communications and missile navigation. RAND's research into the satellite's military usefulness continued into the early 1950s, and the Radio Corporation of America (RCA) signed a contract with the RAND group in June 1952 to study optical systems, recording devices and imagery presentation techniques that might be used on reconnaissance satellites in the future. In July 1953, North American Aviation signed a contract with Wright Field's Communication and Navigation Laboratory to study a pre-orbital guidance system for satellites.[1]

Military space activities up to that time had been concentrated on technological studies and analyses, but the Air Force's Air Research and Development Command (ARDC) redirected the satellite effort toward actual demonstrations of the satellite's major components as part of the Weapon System 117L program in the mid-1950s. The Weapon System 117L program was funded at only 10 percent of the level needed to meet its requirements in 1957 (e.g., $3 million versus $39.1 million), but the Soviets' successful launch of Sputnik I on 4 October 1957 soon forced the U.S. Department of Defense to set higher priorities for the development of military satellite systems. The Department of Defense also created the Advanced Research Projects Agency (ARPA) on 7 February 1958 to supervise all U.S. military space efforts. Though the Navy's VANGUARD satellite project and ARPA's lunar probe project were transferred to the National Aeronautics and Space Administration (NASA) on 1 October 1958, ARPA retained its military satellites, high energy rocket upper stages and its military space exploration programs.[2]

As a spin-off from its Intermediate Range Ballistic Missile (IRBM) development program, most of the Air Force's participation in the Cape's space launch operations in the late 1950s was managed by the WS-315A (THOR) Project Division under the Air Force Ballistic Missile Division's Assistant Commander for Missile Tests. The WS-315A Project Division was redesignated the Space Project Division on 16 November 1959, and it became the Space Projects Division under the 6555th Test Wing on 15 February

1960. The Division supported a total of 10 Air Force-sponsored THOR-ABLE, THOR-ABLE I and THOR-ABLE II space launches at the Cape before the end of 1959. It also supported five THOR-ABLE-STAR missions for the Army, Navy and ARPA in 1960. In the spring of 1960, the Space Projects Division's responsibilities were broadened to include planning for NASA's ATLAS/AGENA-B program at Cape Canaveral.[3]

Following the establishment of Air Force Systems Command (i.e., ARDC's successor) on 1 April 1961, the Air Force's space and missile activities were assigned under two separate intermediate headquarters: the Space Systems Division and the Ballistic Systems Division. The 6555th Test Wing served both intermediate headquarters as their onsite representative at the Cape. Under the 6555th Test Wing's new table of organization, the old Space Projects Division became the Space Projects Branch under the Space Programs Office on 17 April 1961. Under Lieutenant Colonel Harold A. Myers, the 6555th Test Wing's Space Projects Branch focused its attention on satellites and spacecraft being prepared by contractors at Missile Assembly Hangar AA for the TRANSIT, ANNA, RANGER, SAINT and VELA HOTEL projects. Space boosters were monitored by the Space Program Office's other three branches (i.e., ATLAS Boosters, THOR Boosters and BLUE SCOUT) until 25 September 1961, when the THOR Boosters Branch and the Space Projects Branch were combined to form the THOR/TITAN Space Branch. Some of the Space Projects Branch's ATLAS-related functions were transferred to the new ATLAS Space Branch as a result of this reorganization, but the BLUE SCOUT Branch remained intact. After September 25th, the 6555th Test Wing's Deputy for Space Systems accomplished his space mission through the THOR/TITAN Space Branch, the ATLAS Space Branch and the BLUE SCOUT Branch.[4]

Figure 1: THOR-ABLE launch from Pad 17A 7 August 1959

Figure 2: Map of Cape Canaveral Air Force Station

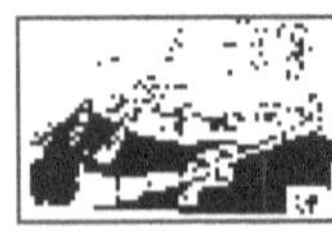

Figure 3: Vicinity Location Map

Two agreements of enormous significance to the 6555th Test Wing's role in the U.S. space program were signed in 1961: 1) the AGENA B Launch Vehicle Program Agreement and 2) the Memorandum of Agreement on Participation of the 6555th Test Wing in the CENTAUR R&D Test Program. The first agreement was signed by NASA Associate Administrator Robert C. Seamans, Jr. and Lieutenant

General Bernard A. Schriever (ARDC Commander) in January and February 1961. It confirmed NASA's intention to pursue the AGENA program through "established USAF Satellite System channels" to take advantage of Air Force capabilities and procedures. NASA remained responsible for its own spacecraft, but it agreed to coordinate its AGENA B requirements through the Air Force Ballistic Missile Division. In effect, this meant the Division ordered NASA's ATLAS boosters and AGENA B upper stages through Air Force contractors (e.g., Convair and Lockheed). Ultimately, NASA had overall responsibility for the countdown and mission, but the 6555th was responsible for the readiness of the entire launch vehicle. The 6555th and the Air Force's contractors were thus very close to the center of NASA's ATLAS/ AGENA B operations.[5]

The second agreement, which was signed on 18 April 1961 by NASA's Dr. Kurt H. Debus and the 6555th Test Wing's Colonel Paul R. Wignall, confirmed that the 6555th would be allowed to exercise launch responsibility for all CENTAUR upper stages used on operational Department of Defense (DOD) missions. The agreement also allowed the 6555th to assign Air Force supervisors to Convair's processing teams while they were working on ATLAS D boosters used for ATLAS/CENTAUR Research and Development (R&D) test flights from the Cape's Launch Complex 36. Though the ATLAS Space Branch acted only as a technical consultant for NASA's CENTAUR development program, it supported NASA's ATLAS booster requirements in accordance with the Seamans/Schriever agreement. The Branch also retained jurisdiction over all military missions involving the ATLAS D and AGENA B as space boosters.[6]

Figure 4: MARINER II Venus Probe

13 August 1962

Figure 5: The AURORA 7 ATLAS/MERCURY mission lifts off
24 May 1962

In 1962, the ATLAS Space Branch supported eight ATLAS and ATLAS/AGENA space missions launched from complexes 12 and 14 (e.g., MARINER I and II, RANGER III, IV and V, and the first three manned orbital ATLAS/MERCURY missions). Air Force contractors prepared all those launch vehicles, but the 6555th started negotiations with its contractors and Air Training Command in September 1962 to establish an all-military "blue suit" ATLAS/AGENA launch team. Though military positions were approved and troops were transferred, the Air Force suddenly dropped the plan and terminated the ATLAS/AGENA blue suit program on 1 January 1964. The SLV-III Division (formerly the ATLAS Space Branch) was reduced to approximately two dozen officers, airmen and civilians by the middle of 1964, and it remained near that level for the balance of the 1960s. The Division's

MERCURY support mission ended following the last MERCURY flight in May 1963, but the unit supported three DOD satellite missions and seven ATLAS/AGENA target vehicle missions for Project GEMINI between October 1963 and mid-November 1966. Under Lt. Colonel Earl B. Essing, the SLV-III Division supported four more classified DOD launches from Launch Complex 13 between 6 August 1968 and 31 August 1970. The Division continued to work in concert with 300 contractor personnel as more classified missions were prepared for launches from Launch Complex 13 in the 1970s.[7]

Figure 6: ATLAS/AGENA Target Vehicle on Pad 14
11 November 1966

Figure 7: BLUE SCOUT JUNIOR at Pad 18
28 January 1965

A blue suit launch capability for ATLAS space vehicles was never developed at Cape Canaveral, but the 6555th retained a limited military launch capability for "guided" BLUE SCOUT and "unguided" BLUE SCOUT JUNIOR space vehicles under the BLUE SCOUT Branch and (later) the SLV-IB Division during the 1960s. The first all-military BLUE SCOUT processing operation was completed in April 1962, and the first all-military launch of a BLUE SCOUT JUNIOR was completed successfully on 30 July 1963. Six more BLUE SCOUT JUNIORs were launched with mixed results before the program was terminated. The SLV-IB Division was disbanded in the last half of 1965, and its personnel were transferred to other agencies under the 6555th.[8]

Figure 8: BLUE SCOUT JUNIOR prelaunch activities
28 January 1965

Figure 9: BLUE SCOUT JUNIOR launch
28 January 1965

As we noted earlier, the 6555th combined its THOR Booster Branch with the Space Projects Branch on 25 September 1961 to form the THOR/TITAN Space Branch. The Wing also established the SLV-V Division on 10 September 1962 to handle the TITAN III launch program separately. (It was renamed the SLV-V/X-20 Division on 1 October 1962.) The THOR/TITAN Branch became the SLV-II/IV

Division on 1 October 1962, but it was split up on 20 May 1963 to form two new divisions: 1) the SLV II Division (for THOR) and 2) the GEMINI Launch Vehicle Division (to oversee the Martin Company's TITAN II/GEMINI operations on Launch Complex 19).[9]

Figure 10: THOR/ASSET on Pad 17B

28 August 1963

With regard to the SLV II Division, launch vehicle contractors supported three Navy satellite flights and 25 NASA THOR/DELTA missions from Launch Complex 17 between the beginning of 1962 and the end of 1965. During much of that period, the SLV II Division concentrated its efforts on the Air Force's two part ASSET (Aerothermodynamic/Elastic Structural Systems Environmental Tests) program which had its first flight from Launch Complex 17 on 18 September 1963. Five more ASSET test flights were launched from Pad 17B between 24 March 1964 and 24 February 1965. Seeing no further military use for THOR facilities at Cape Canaveral, the Air Force directed the 6555th to return them to the Air Force Eastern Test Range in April 1965. The Range subsequently transferred Launch Complex 17 and other THOR/DELTA facilities to NASA's Kennedy Space Center on the condition that the facilities be returned to the Range at the end of NASA's DELTA space launch program. As we shall see in Chapter III, the Cape's DELTA facilities were returned to the Air Force toward the end of 1988.[10]

Concerning the origins of Titan III space operations at the Cape, we need look no farther than the activities of the 6555th Test Wing's SLV-V/X-20 Division, which became simply the "Titan III Division" following cancellation of the DYNA SOAR (X-20) project in December 1963. Within months of its creation in 1962, the SLV/X-20 Division was at the forefront of the 6555th Test Wing's taskforce to supervise construction of TITAN IIIC facilities at the north end of Cape Canaveral. In addition to the Integrate-Transfer-Launch area construction effort, the Division monitored an $819,000 contract with Julian Evans and Associates to modify Launch Complex 20 for four TITAN IIIA missions flown between 1 September 1964 and 7 May 1965. Both TITAN IIIC launch sites (i.e., complexes 40 and 41) had their first launches before the end of 1965, and they continued to support a wide variety of missions for the balance of the decade.[11]

Figure 11: Aerial views of TITAN Integrate-Transfer-Launch (ITL) Area November 1964

Figure 12: TITAN IIIC launch from Pad 40
15 October 1965

On 1 April 1970, the 6555th Aerospace Test Wing was redesignated the 6555th Aerospace Test Group, and it was reassigned to the 6595th Aerospace Test Wing headquartered at Vandenberg Air Force Base, California. That change amounted to a two-fold decline in the 6555th Test Wing's status, but there were good reasons for the action. First, DOD ballistic missile programs at the Cape had become decidedly "Navy Blue" by 1970. U.S. Navy ballistic missile tests constituted more than half of all the major launches on the Eastern Test Range between 1966 and 1972, and the Navy's demand for range services continued without interruption into the 1990s. In sharp contrast to its own ballistic missile efforts of the 1950s and 60s, the Air Force was about to conclude its final ballistic missile test program at the Cape (i. e., the MINUTEMAN III) in December 1970. Second, though the 6555th continued to support space operations from launch complexes 13, 40 and 41, NASA dominated manned space and deep space missions at the Cape. NASA commanded 50 percent of the Eastern Test Range's "range time" as early as 1967, and its status as a major range user was unquestioned. Last but not least, Air Force military space requirements accounted for only 11 percent of the Eastern Test Range's activity, but Air Force space and missile test requirements at Vandenberg accounted for 75 percent of the Western Test Range's workload. With the dramatic shift in Air Force space and missile operations from the Cape to Vandenberg, it was logical to give the 6555th a less prominent role. The 6555th joined the 6595th Space Test Group and the 6595th Missile Test Group as one of three groups under the 6595th Aerospace Test Wing.[12]

From the preceding comments, an observer might conclude that the 6555th Test Group's future looked far less promising than its past in 1970, but events over the next two decades were to prove otherwise. To appreciate the significance of the Test Group's accomplishments in this later period, we first need to examine the organizations, programs and strategies that shaped military space operations at the Cape in the 1970s and 80s. First, there was the Range. As a result of the inactivation of Air Force System Command's National Range Division on 1 February 1972, the Air Force Eastern Test Range (AFETR, pronounced "aff-eater") became the only Air Force Systems Command (AFSC) test range to operate as a separate field command in the 1970s. As such, AFETR had the status of a numbered air force, and it reported directly to AFSC for the next five years. On 1 February 1977, AFETR Headquarters was inactivated, and control of the Range passed to the Space and Missile Test Center (SAMTEC) headquartered at Vandenberg Air Force Base, California. (SAMTEC was the parent organization for the 6595th Aerospace Test Wing, mentioned earlier). Elements of AFETR's old 6550th Air Base Group were reorganized as the 6550th Air Base Wing, which was given host responsibility for Patrick Air Force Base. Rather significantly, the 6550th Wing Commander answered to the AFSC Commander through the latter's Chief of Staff-not the SAMTEC Commander.[13]

Figure 13: POLARIS launch

24 April 1972

Figure 14: POSEIDON launch
15 May 1978

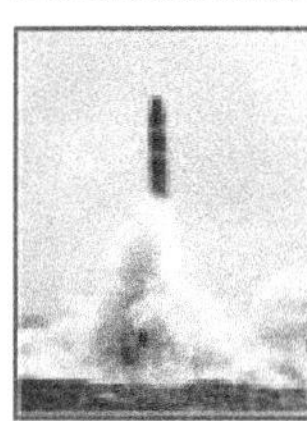

Figure 15: TRIDENT I launch

19 April 1983

Figure 16: TRIDENT II launch

4 December 1989

Figure 17: Final MINUTEMAN III launch from the Cape
14 December 1970

By the summer of 1979, AFSC officials began work on a new organization that would integrate range and launch operations more closely by realigning AFSC elements at Patrick Air Force Base and Vandenberg Air Force Base under consolidated commands. On 1 October 1979, Headquarters SAMTEC was redesignated the Space and Missile Test Organization (SAMTO) and the 6595th Aerospace Test Wing was deactivated. On the same date, The Headquarters, Air Force Eastern Test Range and Headquarters, Western Test Range were both reactivated and redesignated Headquarters Eastern Space and Missile Center (ESMC) and Headquarters Western Space and Missile Center (WSMC) respectively. The 6550th Air Base Wing was deactivated, and its resources were reassigned to the 6550th Air Base Group, which was resurrected under ESMC. The 6595th Shuttle Test Group, 6595th Satellite Test Group and 6595th Missile Test Group were assigned to WSMC, and the 6555th Aerospace Test Group was assigned to ESMC. Both new centers reported to SAMTO, and SAMTO reported to Space Division (formerly the Space and Missile Systems Organization [SAMSO]) headquartered at Los Angeles Air Force Station, California. After nearly twenty years as a tenant unit on the Cape, the 6555th returned to the command relationship it had enjoyed under the Air Force Missile Test Center in the 1950s: it was assigned to a local center (ESMC) which had primary responsibility for Patrick, the Cape and the

Range.[14]

Higher up in the organization, SAMSO and its successors all had the same basic mission: to acquire missile and space systems for the Department of Defense and the U.S. Air Force. Between 1 July 1967 and 30 September 1979, SAMSO was responsible for negotiating and administering contracts for research efforts and space-related hardware. It provided military and civilian space missions with launch vehicles and launch-related operations, and it served as the Department of Defense's Executive Agent for the manned Space Transportation System (a.k.a., the Space Shuttle). From April 1970 through 30 September 1979, SAMTEC served SAMSO by directing all Department of Defense space flights launched from Vandenberg and the Cape. The Space and Missile Test Center and its successor (SAMTO) also evaluated ballistic missile tests on the eastern and western test ranges.[15]

USAF Space Organizations and Programs

The Creation of Air Force Space Command and Transfer of Air Force Space Resources

As space technology evolved in the 1960s and 70s, space systems were applied to the Defense Department's weather, communications, navigation, surveillance and early warning missions. Air Force space operations thus became an indispensable part of the U.S. defense effort by the late 1970s, but the management of space systems was shared among three different organizations (e.g., AFSC, North American Aerospace Defense Command and Strategic Air Command). To redress this fragmentation of its space effort, the Air Staff decided to place space systems operations under a separate major command and create an organization within the Air Staff to supervise the effort. Toward that end, the Directorate of Space Operations was set up under the Air Force's Deputy Chief of Staff for Operations in October 1981, and the Air Force created Air Force Space Command (AFSPACECOM) on 1 September 1982. The transfer of operational space vehicles (i.e., satellites) from one command to another was not expected to happen overnight, but AFSPACECOM would eventually operate the Satellite Early Warning System, the Defense Meteorological Satellite Program (DMSP), the NAVSTAR Global Positioning System (GPS) and various ground surveillance and control systems. Space Division would continue to procure launch vehicles, upper stages and spacecraft for the Air Force even after control of the various satellite systems passed to AFSPACECOM. The organizational link between Space Division and AFSPACECOM would be maintained in the person of the Space Division Commander, who also served as Vice Commander, AFSPACECOM.[16]

As might be expected, one of the new command's greatest challenges in the 1980s was its receipt of control of operational satellite systems from Space Division. The process was very complex and lengthy, and Space Division continued to operate the Air Force Satellite Control Facility (AFSCF) at Sunnyvale, California during the first five years of AFSPACECOM's existence. The Division also pursued the development of the Consolidated Space Operations Center (CSOC) near AFSPACECOM's headquarters at Peterson Air Force Base, Colorado. Following the CSOC ground-breaking ceremony on 17 May 1983, site preparation and construction were soon underway, and the building phase was completed in September 1985. In accordance with a Memorandum of Agreement signed by Space Division and AFSPACECOM in 1982 and supplemented in 1984, CSOC became an AFSPACECOM "service organization" Command owned CSOC's main elements (located in the Satellite Operations Complex), and Space Division transferred most of CSOC's control elements to AFSPACECOM following the inactivation of the AFSCF on 1 October 1987. Air Force Space Command thus became the Air Force's primary agent for satellite operations. While most of the AFSCF's resources were in AFSPACECOM's hands, Space Division's Consolidated Space Test Center (CSTC) still provided telemetry, tracking and control for all military research and development satellites as well as similar services for operational satellites from launch until early orbit.[17]

From the organizational perspective, the transfer of launch operations from Air Force Systems Command to Air Force Space Command was AFSPACECOM's most significant event of the early 1990s. At long last, space operations, from lift-off to satellite deactivation, were assigned to a major *operational* command. Though the transition to full operational status took several years, the path was clearly laid out in AFSPACECOM's Programming Plan 90-2 (Launch Transfer), which was signed by General Ronald W. Yates (for AFSC) and Lt. General Thomas S. Moorman, Jr. (for AFSPACECOM) at the end of August 1990. According to that plan, Phase I of the Launch Transfer began on 1 October 1990. On that date, ESMC, WSMC and their associated range organizations transferred from AFSC to AFSPACECOM along with the 6550th Air Base Group, Patrick Air Force Base, Cape Canaveral Air Force Station and AFSC Hospital Patrick. The aerospace test groups on both coasts remained with AFSC, but two new organizations-the 1st and 2nd Space Launch Squadrons-were activated and assigned to AFSPACECOM on 1 October 1990. The 1st,which was constituted from resources taken from the 6555th Aerospace Test Group at Cape Canaveral Air Force Station, became AFSPACECOM's first DELTA II launch squadron. The 2nd, which was created from resources taken from the 6595th Aerospace Test Group at Vandenberg, became an ATLAS E squadron under WSMC. Though the 6555th Test Group's ATLAS and TITAN resources were not ready to become operational squadrons on October 1st, Air Force Systems Command transformed them into ATLAS II and TITAN IV Combined Test Forces (CTFs) to serve both major commands until such time as they could become operational squadrons under AFSPACECOM. A Combined Test Force for TITAN II/IV operations at Vandenberg was anticipated under Phase II of the Launch Transfer Plan, but, as of this writing, it had not occurred. The TITAN IV CTF (a.k.a. TITAN IV Launch Operations) became the 5th Space Launch Squadron on 14 April 1994.[18]

One more organization must be mentioned before we discuss specific military space programs and strategies: the Strategic Defense Initiative Organization (SDIO). In a speech given in March 1983, President Ronald Reagan proposed the Strategic Defense Initiative (SDI) as a means of rendering nuclear weapons obsolete. (President Reagan subsequently redefined SDI as a research program utilizing new technologies to create effective defenses against ballistic missiles.) The goal of SDI was to strengthen deterrence and provide better security of the United States and its allies. In support of the President's proposal, the Defense Department established the Strategic Defense Initiative Organization in late March 1984, and it named Lt. General James A. Abrahamson as the SDIO's first director. General Abrahamson assumed the Director's post on 16 April 1984. The SDIO established the following five program elements under which individual SDI efforts could be funded and pursued: 1) surveillance, acquisition, tracking and kill assessment (SATKA), 2) directed energy weapons (DEW), 3) kinetic energy weapons (KEW), 4) battle management (command, communications and control), and 5) systems analysis and support programs. The SDIO issued direction for each of those program elements through Work Package Directives (WPDs) to the armed services. The majority of the effort was undertaken by the Air Force and Army. The Air Force managed programs related to boost and mid-course phases of ballistic missile interdiction; the Army managed terminal phase/ground defense programs. Space Division acted as the Air Force's "lead division" for all SDI-related projects except battle management. Battle management was handled by Air Force Systems Command's Electronic Systems Division (ESD). As we shall see in Chapters II and III, SDIO projects and experiments went hand-in-hand with other military space activities at the Cape in the 1980s and early 1990s.[19]

Figure 18:
1st Space Launch Squadron Emblem

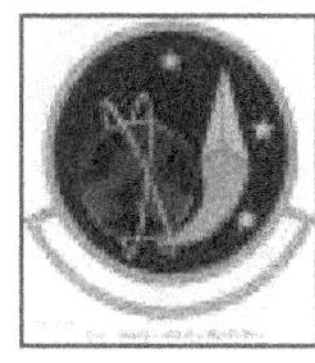

Figure 19:
2nd Space Launch Squadron Emblem

USAF Space Organizations and Programs

Defense Department Involvement in the Space Shuttle Program

Several major space initiatives helped shape the course of the Cape's military space operations after 1970. There were, of course, the Air Force's own programs (e.g., TITAN 34D, TITAN IV, DELTA II, ATLAS II, NAVSTAR GPS, DSCS III, etc.), but the Defense Department also was involved in the nation's manned Space Transportation System (STS) program as both a customer and a supplier of services. The Space Shuttle had a profound effect on the timing and evolution of military space operations on both coasts after 1970.

Following his acceptance of the STS proposal at the end of the 1960s, President Richard M. Nixon authorized NASA to begin developing the Space Shuttle on 5 January 1972. The President authorized a six-year, $5.5 billion development program to have an operational Space Shuttle by the end of the decade. The Shuttle would have a 15x60-foot payload bay, and the orbiter was expected to carry civilian and military payloads weighing up to 65,000 pounds into low-Earth orbit. Since the Shuttle would be expected to accomplish polar missions as well as equatorial missions, Shuttle launch sites on both U.S. coasts were required. In April 1972, the Defense Department agreed with NASA to give the Air Force responsibility for developing a Shuttle launch site and other facilities at Vandenberg Air Force Base to support the Shuttle's west coast missions. Defense had no objection to NASA's decision to convert the Kennedy Space Center's APOLLO launch complexes into Shuttle launch complexes to support the east coast missions. While NASA was responsible for defining the Shuttle's architecture based on inputs from its civilian and military customers, the Defense Department secured an agreement with NASA in October 1973 to allow the Air Force to develop the Interim Upper Stage (IUS) needed to boost Shuttle and unmanned launch vehicle payloads into higher energy orbits. As we shall see below and in Chapter II, the IUS was associated with TITANs as well as Shuttles.[20]

Air Force System Command's Space and Missile Systems Organization (SAMSO) was tasked with the details associated with developing the IUS and acquiring Shuttle facilities at Vandenberg Air Force Base. The development of the IUS got underway first, and it progressed along a separate contracting path from the Vandenberg facility initiative. On 19 August 1974, SAMSO issued a Request For Proposal (RFP) for a nine-month-long IUS system study effort to five major space contractors that had operational upper stages in production (i.e., Martin Marietta, Lockheed, General Dynamics, Boeing and McDonnell Douglas). Four of the five contractors developed IUS concepts based on modifications to existing liquid-propelled upper stages, and Boeing offered an upper stage using solid rocket motors.

Following completion of IUS system studies in 1975, four of the five contractors competed for the IUS validation contract. Based on his assessment of the contractors' IUS system studies and his consultations with the Director, Defense Research and Engineering (DDR&E) and NASA, the Air Force Assistant Secretary for R&D decided to develop the IUS along solid propellant lines. Boeing won the $21.4 million contract on 3 September 1976, and the company proceeded with the 18-month-long validation phase of the IUS effort. One of the primary objectives of that phase of the program was to determine the IUS' optimum size and performance characteristics. The IUS had to be able to lift a 5,000-pound payload into geosynchronous equatorial orbit from a Shuttle cargo bay orbiting 150 nautical miles above Earth. Boeing proceeded on the assumption that the IUS would be 15 feet long and 10 feet in diameter. The basic vehicle would be powered by two solid rocket motors of differing sizes (e.g., a 24,030-pound motor in Stage I and a 7,925-pound motor in Stage II). The larger motor would be rated at approximately 42,000 pounds of thrust, and the smaller motor would provide approximately 17,000 pounds of thrust.[21]

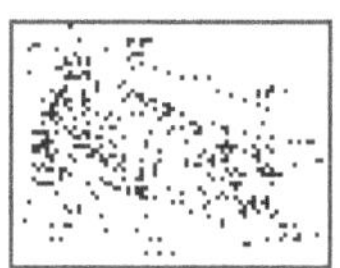

Figure 20: Cutaway profile of Inertial Upper Stage (IUS)

Following the expiration of the Inertial Upper Stage validation contract on 28 February 1978, the Air Force Systems Acquisition Review Council and the Defense Systems Acquisition Review Council both recommended the IUS program be continued to full-scale development. The Deputy Secretary of Defense accepted the recommendations and authorized development of the two-stage IUS and production of the first four IUS vehicles for DOD use. (He also authorized development and production of IUS vehicles for NASA's use, providing NASA funded the effort.) On 18 April 1978, Boeing received a contract valued at $300 million to build the first four DOD inertial upper stages, the first five NASA IUS vehicles, and all their necessary space and ground support equipment. (The purchase was later flip-flopped to five DOD and four NASA vehicles per an interagency agreement in the summer of 1980.) Though serious deficiencies were encountered during IUS solid rocket motor firings in 1979, those problems were resolved and capped during inspections, modifications and test firings over the next two years. Computer software problems and cost overruns also posed other problems, and authorized changes to the contract raised its target cost to $462.2 million by the end of 1980. Despite those technical and financial problems, the IUS program made significant gains in 1981 and 1982. The first IUS flight vehicle completed acceptance testing at Boeing's plant in October 1981. It arrived at the Cape's Solid Motor Assembly Building (SMAB) on 1 March 1982, and it was mated to a TITAN 34D and its Defense Satellite Communications System payload (DSCS II/III) in September 1982. (See Chapter II for further details on that mission.)[22]

With regard to the development of Shuttle facilities at Vandenberg Air Force Base, five major construction projects were on the drawing boards at the beginning of 1978. They consisted of a launch pad, a launch control center, an airfield, a tow route and a mate/demate facility. Martin Marietta completed its design of the Shuttle's ground support systems in August 1978, and the corporation proceeded with a $103 million follow-on contract (awarded in June 1978) to: 1) monitor the design and construction of ground facilities, 2) design ground equipment, 3) develop the software for a computerized launch processing system, 4) plan the activation and operation of ground facilities and 5) plan the logistics support for those facilities. The original plan called for an operational Space Launch Complex 6 (SLC-6) ready to support 20 Shuttle launches per year sometime in 1983. This proved impossible-the program was already facing a $265 million deficit in 1978, and the construction phase of the program had to be lengthened from two years to four years to compensate for the funding problem. While planners remained optimistic about supporting six Shuttle missions per year in 1983, changes in the Shuttle's configuration (i.e., the addition of two strap-on solid rocket motors) required changes in the launch pad's design in 1979, and the pad's initial operating capability (IOC) date slipped to 1984. By October 1981, the IOC had slipped to October 1985, and the total acquisition cost was expected to top $2.5 billion.[23]

Sadly, SLC-6 never saw its first Shuttle launch. The facility was still months away from its latest IOC date when the Challenger disaster grounded the Shuttle program in January 1986. When Air Force Secretary Edward C. Aldridge announced the Defense Department's Space Recovery Plan on 31 July 1986, he sounded the beginning of the end for the Shuttle facility at Vandenberg: the Shuttle program's remaining three orbiters would be kept at the Kennedy Space Center where the backlog in DOD payloads could be reduced as quickly as possible. The Secretary also noted that the Vandenberg Shuttle facility would be completed, but it would be placed in "operational caretaker status." The SLC-6 contractor workforce was to be reduced from 2,100 to approximately 850 people in the fall of 1986 to achieve operational caretaker status, but, following the cancellation of Shuttle tests at Vandenberg in October 1986, the process was accelerated. The Space and Missile Test Organization at Vandenberg ordered the inactivation of the 6595th Shuttle Test Group on 31 January 1987. A small program office was established under WSMC to carry out the Group's remaining tasks, and the launch site was placed on minimum caretaker status on 20 February 1987. Secretary Aldridge directed the Air Force to begin mothballing the facility in May 1988. Large amounts of Vandenberg's Shuttle equipment were transferred to the Kennedy Space Center and its contractors on the east coast. Lesser amounts of equipment were also transferred to other NASA offices, the Navy and Vandenberg's Titan IV program.[24]

Figure 21: Space Launch Complex 6 in 1986

USAF Space Organizations and Programs

Air Force Space Launch Vehicles: SCOUT, THOR, ATLAS and TITAN

Before we discuss the nation's space launch recovery effort, we need to examine the nature of the Air Force's unmanned launch vehicle programs after 1970 and the direction they were headed in the years preceding the Challenger disaster. For its military missions in the 1970s, the Air Force employed the SCOUT, the THOR, the DELTA, the ATLAS and a family of heavy-lift vehicles based on Martin Marietta's TITAN III standard core booster. The smallest of those vehicles was the SCOUT. The SCOUT was a four- or five-stage solid rocket booster manufactured by the Vought Corporation. The basic four-stage SCOUT (SLV-1A) was approximately 72 feet high and 40 inches in diameter at its widest point. It weighed approximately 38,000 pounds. Depending on the configuration used, the SCOUT developed between 103,000 and 130,000 pounds of thrust, and it was capable of lifting small payloads (e.g., 150 to 400 pounds) into low-Earth orbit. The SCOUT was launched by Vought from Vandenberg Air Force Base, Wallops Island (Virginia) and the San Marcos platform off the coast of Kenya. On 19 April 1977, AFSC signed an agreement that transferred technical direction of Vandenberg's SCOUT launches to NASA . The agreement relieved the 6595th Space Test Group of that responsibility for SCOUT.[25]

McDonnell Douglas' THOR and DELTA stood on the next step of the launch vehicle hierarchy. The THOR and DELTA were basically liquid-fueled boosters, and they were sometimes equipped with three, six or nine strap-on solid rocket motors to meet various mission requirements. In the 1970s, the basic THOR launch vehicle (configured with a Rocketdyne RS-27 engine) developed approximately 205,000 pounds of thrust at lift-off. The DELTA launch vehicle (which included a THOR booster as its first stage) was between 112 feet and 116 feet high, and it was eight feet in diameter. Though the DELTA's overall dimensions did not increase appreciably from the 1970s into the late 1980s, the vehicle's lift-off weight and power grew considerably. In 1975, a fully loaded DELTA with nine solid rocket motors weighed 291,000 pounds and provided 518,000 pounds of thrust at lift-off. In 1987, the DELTA used to launch the PALAPA B-2P communications satellite weighed approximately 422,000 pounds at lift-off and provided 718,000 pounds of thrust. Though DELTA missions were managed by NASA on both coasts during the 1970s and most of the 1980s, the Air Force supervised THOR missions launched from Space Launch Complex 10 (West) at Vandenberg until THOR space launches there ceased in 1980.[26]

Figure 22: SCOUT launch from Vandenberg AFB 19 December 1963

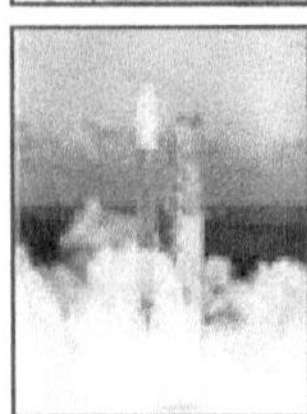

Figure 23: DELTA launch from Complex 17 October 1975

Figure 24: THOR Launch From SLC-10 (West) 14 July 1980

Slightly below the most powerful DELTAs in lift-off power was General Dynamics' liquid-fueled ATLAS/AGENA. The first stage of the vehicle was the ATLAS Standard Launch Vehicle. It measured 82 feet long and 10 feet in diameter. Like the DELTA, the ATLAS burned RP-1 (highly refined kerosene), and it produced approximately 400,000 pounds of thrust at lift-off. The AGENA second stage was 26 feet long and five feet in diameter. It burned Unsymmetrical Dimethylhydrazine (UDMH) and used Inhibited Red Fuming Nitric Acid as an oxidizer. The AGENA was considerably more powerful than the DELTA upper stage (e.g., 16,100 pounds of thrust versus the DELTA's 9,000 pounds of thrust), and this gave the ATLAS/AGENA an edge over the DELTA in boosting heavier payloads into geosynchronous orbits. The combined weight of the ATLAS/AGENA (minus payload) was approximately 292,000 pounds. As noted earlier, the Test Group's SLV-III Division supported classified ATLAS/AGENA missions from Complex 13 during the 1970s. The last of those missions was launched on 6 April 1978.[27]

The 6555th Aerospace Test Group phased out its ATLAS/AGENA operations at the Cape in the spring of 1978, but it still maintained technical oversight for Defense Department payloads launched on NASA's ATLAS/CENTAUR boosters from Launch Complex 36. The first of those payloads was launched on 9 February 1978, and seven more military missions were launched from Complex 36 before NASA returned the site to the Air Force in 1990. The first five payloads were boosted into orbit aboard basic ATLAS/CENTAURs, and the remaining spacecraft flew on newer ATLAS G/CENTAURs. The basic ATLAS/CENTAUR was ten feet in diameter. With its payload and fairing in place, it stood 134 feet tall and weighed approximately 326,000 pounds. The ATLAS/CENTAUR's first stage was only 70 feet long, but it generated approximately 431,000 pounds of thrust at lift-off. Like the ATLAS/AGENA, the ATLAS/CENTAUR's first stage burned RP-1 for fuel. The CENTAUR, on the other hand, burned liquid hydrogen. The basic CENTAUR upper stage was 30 feet long, and it delivered approximately 32,800 pounds of thrust. The ATLAS G/CENTAUR was an improved version of the basic launch vehicle. It was 10 feet in diameter, and it stood 137 feet 7 inches high. It weighed approximately 358,000 pounds (minus spacecraft). The ATLAS G first stage was 72.7 feet long, and it generated approximately 437,500 pounds of thrust. The ATLAS G's CENTAUR upper stage produced approximately 33,000 pounds of thrust at altitude.[28]

Figure 25: Complex 13
1978

Figure 26: ATLAS/CENTAUR launch from Pad 36A
4 May 1979

Figure 27: ATLAS G/CENTAUR launch from Pad 36B
December 1986

Figure 28: TITAN IIIB Space Launch Vehicle

The TITAN IIIB, IIIC and IIID launch vehicles formed the Air Force's family of heavy spacelifters in the 1970s. The "Bs" and "Ds" were launched from Vandenberg Air Force Base, and the "Cs" were launched from the Cape. The TITAN IIIB was the least complicated of the TITAN III operational boosters. It consisted of a TITAN stretched standard core and an AGENA D upper stage. It was used to boost DOD satellites into polar orbits from Vandenberg's Space Launch Complex 4 (West). Nearly five times as powerful as the "B," the TITAN IIID consisted of a standard liquid-fueled core and two solid rocket motors. Each of the solid rocket motors was 85 feet long and 10 feet in diameter. They weighed more than 500,000 pounds apiece and provided a combined thrust of 2,400,000 pounds at lift-off. The TITAN IIID was launched from Space Launch Complex 4 (East), and it boosted heavy payloads (e.g., 24,600 pounds) into low-Earth orbit. In the 1970s, TITAN IIICs were launched only from Launch Complex 40, and they were used to place satellites into very high, geosynchronous equatorial orbits. To accomplish this, the "Cs" were equipped with a standard liquid core, two strap-on solid rocket motors and a liquid-fueled transtage. The TITAN IIIC's core and solid rocket motors were identical to the ones used on the TITAN IIID. The transtage was also ten feet in diameter, but it was 15 feet long. It weighed 28,000 pounds and produced approximately 16,000 pounds of thrust. Depending on the payload fairing that was used, the TITAN IIIC stood between 127 and 150 feet tall.[29]

Figure 29: TITAN IIID launch from SLC-4 (East) Vandenberg AFB

Unlike the THOR and ATLAS, the Air Force intended to hold on to its TITAN III class boosters even after the decision had been made to shift many military space operations away from unmanned launch vehicles (and toward the Space Shuttle) by the early 1980s. With that objective in mind, SAMSO began developing a new variant of the TITAN III called the TITAN 34D in June 1976. The TITAN 34D would replace the TITAN IIIC and TITAN IIID, and it would serve as the primary launch vehicle for some military missions and as a backup system for other payloads scheduled for launch aboard the Space Shuttle. Though heavy-lift missions from Vandenberg might only require Titan 34Ds equipped with radio guidance systems, the TITAN 34Ds launched from the Cape would each be equipped with an IUS to meet more complicated flight plans. (Anticipating that requirement, the IUS was developed for both the TITAN 34D and the Space Shuttle to allow greater flexibility and economies of scale.) In September 1978, SAMSO issued a request for proposal for its first three TITAN 34Ds with an option to buy two more in May 1980. The vehicle's specifications called for a slightly longer 110-foot-long Stage I/Stage II core and 90-foot-long solid rocket motors. All booster diameters remained unchanged from the TITAN IIIC/D configuration, but lift-off power improved to 2,800,000 pounds of thrust. Stage I's thrust increased to approximately 530,000 pounds. The Air Force accepted its first three TITAN 34Ds in August 1980. Two more 34Ds were accepted in November and December 1980. The last TITAN IIIC lifted off Complex 40 in March 1982, followed by the first TITAN 34D on 30 October 1982. The last TITAN IIID lifted off Space Launch Complex 4 (East) on 17 November 1982.[30]

Figure 30: TITAN IIIC launch from Complex 40 14 December 1975

Figure 31: First TITAN 34D launch from the Cape 30 October 1982

USAF Space Organizations and Programs

Early Space Shuttle Flights

The Air Force hoped the TITAN 34D and Space Shuttle would complement each other on heavy space operations in the early 1980s, but the Shuttle's slow development was a cause for some concern. Though the Space Shuttle would eventually become the centerpiece of America's space effort, the vehicle was plagued with thermal tile and engine problems late in its development. Those problems continued to delay the Shuttle's debut, and they contributed to the Air Force's decision to hold on to the "mixed fleet" approach to military space operations several years before the Challenger disaster. At the beginning of 1979, as Rockwell International struggled to install thermal protective tiles on the first Shuttle orbiter (Columbia) before its ferry flight to the Kennedy Space Center, the corporation's Rocketdyne Division had to retest the orbiter's main engines to verify the successful redesign of the engines' main oxidizer valves. That testing was successful, but a fuel valve failed during another engine test in July 1979. Engine testing resumed in October 1979, and it continued into the middle of March 1980. In the meantime, Columbia had been ferried to the Kennedy Space Center (KSC) toward the end of March 1979. The new orbiter still lacked 10,000 of its 34,000 thermal protective tiles, and those tiles were painstakingly installed by a special NASA team at the rate of about 600 per week through the summer of 1979. Problems with tiles and engines continued to delay NASA's first launch of Columbia in 1980, and that mission was eventually postponed to March 1981. Since the first four Shuttle flights would be test flights, the first operational mission was not expected before September 1982. Consequently, some payloads scheduled for Shuttle missions had to be delayed or transferred to more unmanned launch vehicles. The Space and Missile Systems Organization contracted six more TITAN 34Ds with Martin Marietta in the summer of 1979, and Space Division (i.e., SAMSO's successor) placed another order for five TITAN 34Ds with Martin in November 1980.[31]

While work on Columbia continued at KSC, the U.S. Army Corps of Engineers marked the completion of IUS processing facility modifications to the east end of the Cape's Solid Motor Assembly Building (SMAB) in September 1980. The west low bay area of the SMAB was set aside to support IUS/Shuttle payload integration, and a $15,900,150 contract for the construction of the Shuttle Payload Integration Facility (SPIF) in that area was awarded to Algernon Blair Corporation of Atlanta, Georgia on 10 July 1981. Construction on the SPIF began on 4 August 1981, and it was completed on 15 February 1984. A separate contract to operate the SPIF was awarded to McDonnell Douglas Technical Services Company on 22 April 1982, and, following installation of equipment and completion of the facility in February 1984, the SPIF was declared fully operational. The SPIF was designed as a processing facility for military payloads, but Space Division agreed to let NASA use the SPIF for civilian payloads on a "case-by-case" basis as long as the agency paid for the processing work and complied with the Defense

Department's security requirements.[32]

Figure 32: Shuttle COLUMBIA November 1980

Ten Shuttle missions had been launched from KSC by the time the SPIF went into operation, and those flights set the tone for the Air Force's launch vehicle strategy in the mid-1980s. The schedule for the first of these missions (i.e., Columbia, March 1981) had been extremely tight. As events unfolded, three different problems arose to delay that flight by about one month. Nevertheless, with veteran astronaut John W. Young in command and Navy Captain Robert L. Crippen as pilot, Columbia lifted off Pad 39A on 12 April 1981 at 7:00 a.m. Eastern Standard Time. For that 54-hour mission, Columbia carried instruments to measure orbiter systems performance, but no payloads were carried in the orbiter's payload bay. Tests of Columbia's space radiators, maneuvering and attitude thrusters, computers, avionics systems and protective tiles all went well, and Columbia's crew landed the orbiter at Edwards Air Force Base on April 14th. Columbia's second mission was scheduled to be launched from Pad 39A on October 9th, but its actual lift-off occurred there on 12 November 1981. The third Shuttle test mission was launched from Pad 39A on 22 March 1982, and the last of Columbia's test flights was launched from Pad 39A on 27 June 1982. The mission on June 27th featured the first Shuttle launch of a Defense Department payload, and it generated several Air Force "lessons learned" reports designed to enhance the future of military space operations aboard the Shuttle.[33]

The next six Shuttle flights featured a variety of payloads, problems and mission results. Columbia's STS-5 flight was the first operational mission of the Shuttle program. Its primary objective was to launch two communications satellites (Satellite Business Systems' SBS-3 and TELESAT CANADA's ANIK C-3). The countdown went well on 12 November 1982, and both satellites were deployed successfully on the first and second days of the mission. The sixth Shuttle mission (STS-6) was Challenger's maiden flight, and it featured the first extravehicular activity (i.e., spacewalk) in the history of the Shuttle program. Preparations for Challenger's first flight were anything but uneventful: the launch scheduled for 20 January 1983 was delayed when a hydrogen leak was detected in Main Engine Number 1 in mid-December 1982; the launch was postponed for two months, but a replacement engine also developed a leak, and a second replacement engine had to be checked out and shipped to KSC in late February 1983. Leaks were detected in another component of Main Engine Number 2, and all three of Challenger's main engines were pulled for repairs at the end of February. The launch was rescheduled for March 26th, but contamination was detected in the payload bay area in early March, so the launch was rescheduled for early April 1983. Under the command of Paul J. Weitz, Challenger finally lifted off on its five-day mission from Pad 39A at 1830:00 Greenwich Mean Time on April 4th. The first in a series of Tracking and Data Relay System (TDRS) satellites was deployed on Challenger's first day in orbit, but the IUS used to propel the payload into geosynchronous orbit had a premature cutoff, and several months of altitude correction by the satellite's thrusters were required before the satellite entered normal service.[34]

Figure 33: Launch of Shuttle COLUMBIA 27 June 1982

Two Hughes HS-376 series satellites were deployed successfully during Challenger's second mission (STS-7) in mid-June 1983, but ground control checkout problems with the TDRS satellite deployed during STS-6 delayed support long enough to push Challenger's third mission from August 20th to the end of the month. That flight (STS-8) was launched at 0632:00 Greenwich Mean Time on 30 August 1983, and it proved to be the smoothest Shuttle mission up to that time. An Indian National Satellite (INSAT-1B) was deployed from the orbiter on Day Two of the mission, and TDRS-A communications tests went well. Extensive testing of the orbiter's Remote Manipulator System (RMS) was successful.[35]

The ninth Shuttle mission (STS-9) was Columbia's only flight during 1983, and it was conducted from November 28th through December 8th. It was commanded by John Young and piloted by Brewster H. Shaw, Jr. The flight was primarily a "shakedown" mission to test the compatibility of Space Lab systems with orbiter systems, but secondary objectives included investigations into atmospheric physics, plasma physics, astronomy, material and life sciences and Earth observations. At least 70 experiments in those disciplines were performed during STS-9. The Space Lab science crew consisted of Dr. Byron K. Lichtenberg and Dr. Ulf Merbold. Dr. Owen K. Garriott and Dr. Robert A. Parker served as mission specialists. In circumstances reminiscent of STS-8, Columbia's mission was rescheduled to late October 1983 due to delays in TDRS-A verification testing. A further delay occurred in mid-October when Columbia had to be rolled back to the Vehicle Assembly Building (VAB) at KSC and demated from its external tank so a suspicious exit nozzle on the right-hand solid rocket booster could be replaced. The procedure required a complete restack of the orbiter/booster assembly, and it delayed the launch until late November. The countdown and launch on November 28th was uneventful, and on-orbit operations went well. Unfortunately, a significant problem surfaced during deorbit operations: two of the orbiter's five General Purpose Computers (GPCs) broke down, and only one of the two computers could be reinitiated. The Flight Director waved off the planned deorbit, and the landing was rescheduled to give officials time to study the computer problem. Columbia finally made a lakebed runway landing at Edwards about seven hours and 48 minutes later than scheduled.[36]

Under the command of Vance C. Brand, Robert L. Gibson piloted Challenger on the tenth Shuttle mission (41-B) from the 3rd through the 11th of February 1984. Bruce McCandless II, Ronald E. McNair and Robert L. Stewart served as mission specialists on the flight. The mission included the first untethered spacewalks by McCandless and Stewart, the first use of the Manned Maneuvering Unit, and the first Shuttle landing at KSC. The orbiter experienced very few problems during the mission, but Western Union's WESTAR VI communications satellite and Indonesia's PALAPA B-2 spacecraft both wound up in useless low-Earth orbits after their Payload Assist Modules (PAMs) malfunctioned following deployment. (Both satellites were retrieved during Shuttle mission 51-A in November 1984.) Challenger landed without incident on February 11th. (footnote #37)

USAF Space Organizations and Programs

Origins of the TITAN IV Program

The Space Transportation System accumulated an impressive number of "firsts" during its first three years of test flights and operations, but the Air Force was concerned about the Shuttle's slower-than-anticipated turnaround time and the impact it would have on the Defense Department's launch schedule in the long term. The Air Force had already received nine TITAN 34Ds from Martin Marietta by September 1981, and it would receive four more 34Ds to meet vital military mission requirements and fill the anticipated Shuttle/DOD mission gap. As Shuttle delays continued, this was clearly not enough. In testimony before Congress in March and April 1984, Air Force officials argued that a new, more powerful, commercially-procured expendable launch vehicle (CELV) ought to be purchased to support those military spacecraft programs that were too vital to become entirely dependent on the Shuttle, particularly in time of war. Put simply, the Defense Department and the Air Force believed that dependency on the Shuttle for those missions was foolish, since any catastrophic accident could ground the Shuttle fleet for unacceptable lengths of time. Although Congress appeared willing to accept CELVs in the interest of national security, it was not willing to accept the commercial contract arrangements proposed by the Air Force. Instead, the CELVs (now called Complementary Expendable Launch Vehicles) would be funded incrementally, beginning with about $5 million in new authorizations and $30 million in reprogrammed funds in 1985. This was the beginning of the TITAN IV program, which reaffirmed the Air Force's long-term commitment to expendable launch vehicles into the 1990s.[38]

The Air Force had already reprogrammed $1 million for the CELV's concept definition in January 1984, and, now that Congress accepted the proposal, Space Division prepared to award a contract for the initial study and development of the CELV based on replies to a revised Request For Proposal (RFP) published in July 1984. Though NASA protested the CELV action, the agency grudgingly submitted its own candidate: a TITAN III liquid core and CENTAUR upper stage supported by three solid rocket boosters. Convair's candidate was a greatly enlarged ATLAS (e.g., 200 inches in diameter) equipped with five new liquid rocket engines, four strap-on solid rocket motors and a CENTAUR upper stage. The winning proposal was Martin Marietta's TITAN 34D7 (later known as the TITAN IV). It consisted of a lengthened TITAN core, a Shuttle-configured CENTAUR G upper stage and two seven-segment (versus the 34D's five-and-one-half segment) solid rocket motors. Martin Marietta believed their new vehicle would be capable of placing a 10,000-pound payload into geosynchronous orbit. Space Division awarded the initial $5,100,000 contract to Martin Marietta on 28 February 1985. Though the initial contract was very small, great significance was attached to its options for fabrication and delivery of complete TITAN IV vehicles. At the time the contract was awarded, those options were valued at $2,095,800,000.[39]

Before the Challenger disaster in January 1986, the Air Force planned to purchase approximately ten TITAN IVs. The Cape anticipated two TITAN IV launches per year beginning in October 1988, and the Air Force intended to launch its remaining ATLAS, TITAN 34D and converted TITAN II missile/space boosters as well. Clearly, the Air Force was committed to unmanned launch operations in the 1980s, and the service maintained that commitment when faced with two TITAN 34D launch mishaps in 1985 and 1986. On 28 August 1985, a TITAN 34D launch failure at Vandenberg led to an extensive accident investigation and a temporary suspension of TITAN 34D launch operations. Following the Challenger tragedy in January, another TITAN 34D was launched from Vandenberg on 18 April 1986 with disastrous results. That mishap effectively grounded TITAN 34D operations on both coasts until an aggressive recovery program could be implemented to ensure the TITAN's reliability. The next TITAN 34D was launched successfully from Vandenberg on 26 October 1987.[40]

Following the tragic and costly launch failures of 1986, Air Force Secretary Edward C. Aldridge presented the Defense Department's plan for space launch recovery on 31 July 1986. Secretary Aldridge had championed the CELV concept earlier, and he now unveiled a plan that called for the development of a "mixed fleet" of expendable launch vehicles. (The Air Force had never completely given up that concept in the first place, and the plan confirmed the wisdom of spreading launch risks among several dissimilar launch vehicle systems.) Among his recommendations, the Secretary proposed expanding the scope of TITAN II and TITAN IV space operations and initiating a new medium launch vehicle program. After considerable discussion and thought, Congress and the Reagan Administration agreed to the plan. The plan implemented a shift in U.S. space policy away from the Shuttle for many military missions, and it heralded a whole new era of opportunities for the unmanned booster companies.[41]

Figure 34: TITAN 34D launch failure 18 April 1986

USAF Space Organizations and Programs

Development of the ATLAS II and DELTA II Launch Vehicles and the TITAN IV/CENTAUR Upper Stage

After the Challenger disaster and the second TITAN 34D mishap, the Air Force awarded Martin Marietta a $1,558,000,000 contract for 13 additional TITAN IVs. The Air Force also planned to buy from 12 to 20 new boosters based on a "yet-to-be-defined" Medium Launch Vehicle (MLV) design. The MLVs were needed to launch Global Positioning System (GPS) NAVSTAR Block II satellites that were rescheduled from the Shuttle launch manifest following Challenger's ill-starred mission. On 1 August 1986, Space Division awarded four six-month R&D contracts worth $5 million apiece to McDonnell Douglas, Martin Marietta, General Dynamics and Hughes Aircraft to develop the initial MLV concept. In an effort to meet an initial MLV/NAVSTAR launch date in January 1989, Space Division also released the RFP for the MLV production/launch contract shortly thereafter. Hughes, General Dynamics, Martin Marietta and McDonnell Douglas delivered their proposals for the new contract in late October 1986, and Space Division awarded the MLV contract to McDonnell Douglas Astronautics Company on 21 January 1987. Under the basic contract (valued at $316,504,000), McDonnell Douglas agreed to deliver and launch seven upgraded DELTA space boosters called DELTA IIs. The contract had options for the production and launch of 13 more DELTA IIs, bringing the potential value of the entire contract up to approximately $669 million.[42]

Two versions of the DELTA II were built under the contract and its options: the basic Model 6925 and the improved Model 7925. Both models featured an 85.7-foot-long first stage (versus the old DELTA's 74-foot-long first stage), a 19.6-foot-long second stage, wider payload fairings (9.5 feet versus 8 feet for the DELTA) and more powerful Morton Thiokol Corporation Castor IVA solid-propellant rocket motors. Fully assembled, the DELTA IIs stood approximately 130 feet tall. Both models were designed to boost NAVSTAR Block II satellites into 10,898-nautical-mile-high orbits, but the less powerful Model 6925 was limited to payloads around 1,850 pounds. As a result of this limitation, Model 6925s were scheduled to be used on only the first nine NAVSTAR Block II missions launched from the Cape. The more powerful Model 7925s would boost more advanced Block II payloads weighing up to 2,100 pounds apiece. Though the DELTA II program office estimated the first launch might occur as early as October 1988, a slow startup in McDonnell Douglas' production facility at Pueblo, Colorado pushed the launch date back to its original estimate of January 1989. The first DELTA II/NAVSTAR mission was launched from Pad 17A on 14 February 1989. The Model 6925 used for that flight performed well, and the mission was successful.[43]

Figure 35: First DELTA II/NAVSTAR launch 14 February 1989

The Challenger disaster had profound and, in some instances, salutary effects on the launch community, but it also contributed to NASA's decision to give up its CENTAUR G' (upper stage) project . Less than two weeks before the fateful mission, NASA's Associate Administrator for Space Flight (Rear Admiral Richard H. Truly) had been pressing his people to resolve "open items" pertaining to the CENTAUR. Following the tragedy on January 28th, KSC's Safety Officer refused to approve advanced processing of the first CENTAUR in February 1986. He cited "insufficient verification of hazard controls" from the Lewis Research Center and the CENTAUR's developer, General Dynamics. In March, NASA Chief Engineer Milton A. Silveira informed Rear Admiral Truly that there was "a significant degree of concern" about Shuttle/CENTAUR safety. Admittedly, there were also serious concerns over the CENTAUR's cost overruns (e.g., $100 million in 1985 and approximately $170 million for safety corrections in 1986), but the Challenger disaster forced NASA to take a second hard look at the safety aspects of the liquid oxygen/hydrogen-fueled CENTAUR. In the safety-sensitive atmosphere of the post-Challenger era, the CENTAUR would never see the inside of a Shuttle cargo bay. On 23 April 1986, Rear Admiral Truly ordered a halt to all work on the CENTAUR; after further discussions with his safety experts in May and June, Truly prudently decided to cancel the Shuttle/CENTAUR project on 19 June 1986.[44]

Cancellation of the Shuttle/CENTAUR project forced the Air Force to replan its TITAN IV/CENTAUR program, but there were opportunities as well as misfortunes arising from this development. The TITAN IV/CENTAUR G had been managed by Space Division through its TITAN IV contract with Martin Marietta since February 1985. A stop work order at the inception of the program had already delayed the start of the TITAN IV/CENTAUR G by about a year, and termination of the Shuttle/CENTAUR program caused further delays as General Dynamics regrouped its CENTAUR effort. The elimination of the Shuttle/CENTAUR program ended the Air Force's "piggyback ride" on NASA's CENTAUR G' technology, but it also disentangled the G's development from the G' and forced the Air Force to look to its TITAN IVs as the only path into space for payloads requiring the CENTAUR upper stage. The CENTAUR's avionics and weight problems came clearly into focus under this pressure, and the Air Force conducted a thorough structural review of the CENTAUR's design in June 1986 to ensure the upper stage could withstand a TITAN IV lift-off. In July 1986, the CENTAUR program office encouraged General Dynamics to "consider all realistic enhancements" for the CENTAUR's major subsystems. There was also an increased demand for high fidelity CENTAUR data due to the increased number of TITAN IV/CENTAUR customers.[45]

Like the DELTA program, the ATLAS/CENTAUR program got a new lease on life under military and commercial auspices following the Challenger disaster. Unfortunately, ATLAS II/CENTAUR operations did not begin as quickly as the DELTA II's operations: the transition was hampered by a two-year delay in the ATLAS/CENTAUR-68 mission due to a ruptured CENTAUR upper stage. NASA sponsored its final ATLAS/CENTAUR mission from Pad 36B on 25 September 1989, and Complex 36 was transferred back to the Air Force in January 1990. In the meantime, the Air Force transferred its Defense Satellite Communications System (DSCS) III payloads and several other military payloads from the Shuttle's manifest to a launch schedule for a new medium launch vehicle, advertised as the MLV II. (The new vehicle was needed to boost payloads weighing as much as 4,900 pounds into low-Earth orbit from Vandenberg and up to 5,800 pounds into geosynchronous orbit from the Cape.) The MLV II would be contracted out as a "commercial launch service acquisition." This meant that the contractor had to agree to provide all services needed to test, process, integrate and launch the vehicle. McDonnell Douglas and General Dynamics dominated the MLV II competition in early 1988, and General Dynamics Space Systems Division won the contract on 16 June 1988. The MLV II would be based on an elongated version of General Dynamics' ATLAS/CENTAUR vehicle. The basic contract called for the production and launch of two ATLAS II/CENTAURs, and there were options for nine more vehicles and launches. Together with all its options, the ATLAS II/CENTAUR contract was valued at more than $500 million. The contract was placed on a Firm-Fixed-Price basis, and the government expected each ATLAS II/CENTAUR launch to cost no more than $40 million under that arrangement (i.e., $100 million less apiece than launches aboard the Shuttle). The first military ATLAS II/CENTAUR was launched from Pad 36A on 11 February 1992, and it carried an operational DSCS III spacecraft into orbit. General Dynamics also fielded commercial versions of its new booster (e.g., ATLAS II/CENTAUR and ATLAS IIA/CENTAUR). The first commercial ATLAS II/CENTAUR was launched from Pad 36B on 7 December 1991, and the first commercial ATLAS IIA/CENTAUR lifted off the same site on 10 June 1992. The ATLAS II/CENTAUR was praised as the "final link" in the Air Force's space launch recovery program.[46]

USAF Space Organizations and Programs

Space Shuttle Support of Military Payloads

While the Space Shuttle did not become the all-encompassing spacelifter its planners hoped for, it would be extremely unfair to ignore the Shuttle's contribution to the military space effort in the 1980s and early 1990s. Following the first DOD/Shuttle mission launched in June 1982, the first two Defense Department payloads processed through the SPIF were launched on two mixed military/civilian missions on 30 August and 8 November 1984. Both of those military payloads were SYNCOM IV communications satellites destined to replace aging FLTSATCOM satellites. Two more SYNCOM IVs were launched on mixed missions in April and August 1985, and two Shuttle missions in January and October 1985 were completely devoted to classified Defense Department payloads. Military payloads continued to be launched on the Shuttle after the "return to flight" on 29 September 1988. Shuttle/DOD missions were launched in December 1988, August 1989, November 1989, February 1990, November 1990, April 1991 and November 1991. In addition to those mostly classified missions, the fifth SYNCOM IV satellite was launched on Columbia's Long Duration Exposure Facility retrieval mission in January 1990. Taken together, military space missions accounted for part or all of 14 out of 37 Shuttle flights launched from the Cape between August 1984 and July 1992. Unquestionably, this represented a major contribution to the U.S. military space effort in the 1980s and early 1990s.[47]

USAF Space Organizations and Programs

U.S. and Soviet Military Space Competition in the 1970s and 1980s

A wide variety of military spacecraft were launched from the Cape during the 1970s, 80s and early 90s. Some of them have already been mentioned in connection with launch vehicle programs presented in this chapter, and most of them will be addressed in somewhat greater detail (where security guidelines allow) as individual military missions in the next two chapters. Before we move on, we should note the operative principles that put all those individual efforts into a growing national military space strategy in the 1980s.

It is safe to say that the U.S. and Soviet "space race" of the 1960s had already given way to a very low-key, sophisticated competition for military superiority in space by the late 1970s. Like the five fingers of a hand, the U.S. and the Soviet Union measured their military space capabilities in terms of: 1) communications, 2) navigation, 3) reconnaissance/early warning/weather surveillance, 4) offensive systems (i.e. ballistic missiles) and 5) defensive systems. American political leaders saw military space systems mature in the 1970s, but they were soon confronted with Central Intelligence Agency (CIA) and National Security Agency (NSA) reports that suggested the Soviet Union had narrowed the "technology gap" between itself and the United States. The Reagan Administration maintained that Soviet defense outlays exceeded American defense spending by 20 percent in 1972, by 55 percent in 1976 and by 45 percent in 1981. Throughout that period, the Soviet Union had reportedly spent 10 percent more than the U.S. in the crucial areas of research, development, test and evaluation. (In the area of defensive/offensive space capabilities, some officials believed years of Soviet research into the principles of directed energy might yield experimental particle beam and laser "devices" by the mid-1980s.) Though many Democrats and some Republican congressmen challenged those claims, popular support for a stronger national defense grew. Thus, despite program cost overruns and launch failures, three factors-politics, Soviet military competition, and rapidly changing technology-coalesced to commit the United States to deploy more capable military space systems and launch more sophisticated space experiments in the late 1980s. In the next two chapters, we will examine those efforts, and we will look at the field agencies and contractors who made those individual launch operations possible.[48]

Chapter One Footnotes

Cape Canaveral
As the reader may already be aware, Cape Canaveral was not always known as "Cape Canaveral." Under Presidential Executive Order No.11129 (29 November 1963), Air Force facilities on Cape Canaveral and NASA's holdings on Merritt Island were designated as the John F. Kennedy Space Center. In line with this order, the Air Force changed the name of its reservation on Cape Canaveral from Cape Canaveral Missile Test Annex to Cape Kennedy Air Force Station on 22 January 1964. The Cape and the Air Force Station were officially renamed Cape Canaveral and Cape Canaveral Air Force Station on 1 April 1974. To avoid confusing the reader, we will refrain from switching back and forth between "Cape Kennedy" and "Cape Canaveral." We will refer to both the Cape and the Air Force Station as simply "the Cape."

THOR-ABLE, THOR-ABLE I and THOR-ABLE II - THOR-ABLE-STAR
The THOR ballistic missile was used as the first stage for each of those launch vehicles. The THOR weighed 110,400 pounds, and it was 62.5 feet long and 8 feet in diameter. It was propelled by a single liquid-fueled rocket motor rated at between 135,000 pounds and 150,000 pounds of thrust. The ABLE second stage was an Aerojet-General booster rated at 7,700 pounds of thrust. The ABLE I added the Allegheny Ballistic Laboratory's 2,450-pound-thrust solid rocket as the third stage to the ABLE second stage. The THOR ABLE II consisted of a THOR first stage and a modified Aerojet-General 10-40 second stage. Aerojet-General's ABLE-STAR upper stage was designed to boost a 1,000-pound payload into a 300-mile orbit.

Space Programs Office
The Space Programs Office was renamed the Office of the Deputy for Space Systems on 25 September 1961.

three branches (i.e., ATLAS Boosters, THOR Boosters and BLUE SCOUT)
The THOR and ATLAS booster branches were outgrowths of the Air Force's ballistic missile programs. The BLUE SCOUT Branch evolved from the 6555th's involvement in Aeroneutronic's small solid rocket experimental launch operations on Launch Complex 18 at the Cape.

ATLAS D and AGENA B
The ATLAS D space booster was essentially a modified ATLAS Intercontinental Ballistic Missile (ICBM). Like the ATLAS D series missile, the ATLAS space booster was 75 feet long and 10 feet in diameter. It was a kerosene-fueled vehicle powered by two (first-stage) 154,000-pound-thrust Rocketdyne vernier booster engines and a 57,000-pound-thrust (half stage) sustainer engine. The

AGENA B upper stage was 21.6 feet long and five feet in diameter. It used Unsymmetrical Dimethylhydrazine (UDMH) for fuel and Inhibited Red Fuming Nitric Acid (IRFNA) as an oxidizer. The ATLAS D was used to launch MERCURY capsules, and the ATLAS/AGENA B combination was used to launch other spacecraft. The combined weight of the ATLAS/AGENA-B vehicle (minus payload) was approximately 292,000 pounds.

BLUE SCOUT
The BLUE SCOUT I launch vehicle consisted of an Aerojet-General solid rocket, a Thiokol TX-33 solid rocket and an Allegheny Ballistic Laboratory ABL-X254 solid rocket. The SCOUT and the BLUE SCOUT II both included those rocket stages, plus an Allegheny Ballistic Laboratory ABL-X248 rocket. The BLUE SCOUT JUNIOR consisted of a TX-33, an ABL-X254, an Aerojet-General AJ 10-41 rocket motor and the NOTS 100A solid rocket. Following six launches in 1961, BLUE SCOUT operations at Launch Complex 18 were scaled back drastically but not eliminated. Toward the end of 1961, Aeroneutronic began providing only limited assistance to the BLUE SCOUT Branch via a Letter Contract.

TITAN III
TITAN IIIA and TITAN IIIC launch vehicles were configured around a modified TITAN II ICBM first stage as their first stage core. That core stage was rated at 430,000 pounds of thrust at sea level, and it provided the Titan IIIA with all its power at lift-off. Two 10-foot diameter solid rocket motors were attached to the basic "A" configuration to make the TITAN IIIC, and those five-segment solid rockets developed 2,314,000 pounds of thrust-all the power the 1,300,000-pound TITAN IIIC needed to lift itself off the pad. The TITAN IIIC's first stage core fired at an altitude of 28 nautical miles, later in the flight. Both vehicles employed a liquid-fueled second stage (rated at 100,000 pounds of thrust) and a small, pressure-fed transtage (rated at 16,000 pounds of thrust) to place their payloads into orbit.

TITAN IIIC facilities
The Cape's TITAN IIIC construction program began in earnest on 24 November 1962 after a $4.6 million contract was awarded to the Atlantic Gulf and Pacific Company to prepare sites for launch complexes 40 and 41 at the north end of Cape Canaveral. Dredging operations in the shallows of the Banana River were underway by February 1963 to move 6.5 million cubic yards of landfill from the river to the Integrate-Transfer-Launch (ITL) sites. The contract for the TITAN IIIC launch complexes was awarded to C.H. Leavell and Peter Kiewit & Sons on 13 June 1963, and it was completed in 1965 for approximately $17 million. Most of the other ITL facilities were grouped under a $26.8 million contract awarded to the firm of Paul Hardeman and Morrison-Knudsen on 30 July 1963. That contract was completed on 16 April 1965.

wide variety of missions
The missions included: 1) the successful release of seven Initial Defense Communications Satellite Program (IDCSP) satellites and one gravity gradient satellite in June 1966, 2) the successful orbit of a modified GEMINI spacecraft plus three secondary satellites in November 1966, 3) the successful release of eight IDSCP satellites in January 1967, 4) the launch of three IDSCP

satellites, the LES-5 satellite and two other payloads in July 1967, 5) the successful orbit of eight IDSCP satellites in June 1968, 6) the launch of the LES-6 communications satellite and three scientific satellites into various orbits in September 1968, 7) the orbit of a 1,600-pound Air Force communications satellite in February 1969, 8) the launch of two VELA and three experimental satellites in May 1969, 9) the successful orbit of two more VELA satellites in April 1970 and 10) the launch of a classified DOD payload in November 1970.

only Air Force Systems Command (AFSC) test range to operate as a separate field command
The Air Force Western Test Range (AFWTR) was under the control of the Space and Missile Test Center (SAMTEC), and the Eglin Gulf Test Range was under the control of the Armament Development and Test Center (ADTC).

SAMSO
SAMSO's space and missile functions were reorganized under two new entities (Space Division and the Ballistic Missile Office) on 1 October 1979. Space Division became Space Systems Division on 15 March 1989. The Ballistic Missile Office was redesignated the Ballistic Systems Division on 15 March 1989, and it became the Ballistic Missile Organization under Space Systems Division on 5 May 1990.

Space Division
Space Division remained responsible for managing the research and development, testing, procurement and launch of most of the nation's military space systems. Acquisition of space assets thus remained an AFSC and Space Division function.

AFSCF
The AFSCF had been set up at Sunnyvale to perform three important functions: 1) on-orbit checkout of experimental and operational satellites, 2) trouble-shooting problems with satellites, and 3) normal satellite control operations. Seven remote tracking stations were added to create the Air Force Satellite Control Network (AFSCN) in the 1960s. As satellite constellations became larger and more varied, some normal control operations were transferred to other satellite mission control systems (e.g., ground stations operated by Air Defense Command and, later, by SAC), but the AFSCF continued to perform tasks 1 and 2 and control Remote Tracking System (RTS) facilities.

The 1st
The 1st was assigned to ESMC on 1 October 1990, and it was placed under the 45th Operations Group upon activation of the 45th Space Wing on 12 November 1991.

6555th
One hundred and seventy-five Test Group personnel were transferred "on paper" from the 6555th Aerospace Test Group to the 1st Space Launch Squadron and the ATLAS II and Titan IV CTFs in October 1990. Sixty-six personnel remained with the 6555th formally after 1 October 1990, but

that number dwindled to around thirty-six people by the end of 1991. Following the inactivation of ESMC and the activation of 45th Space Wing on 12 November 1991, the 1st Space Launch Squadron was reassigned from ESMC to the 45th Operations Group. Both CTFs were placed under the 45th Operations Support Squadron (45th Operations Group) until they could become fully operational squadrons in their own right. Most of the people assigned to the new organizations performed many of the same tasks they had before the transfer, but their training and reporting procedures became increasingly operational in nature. The ATLAS II CTF became the 3rd Space Launch Squadron after the second military ATLAS II/CENTAUR launch on 2 July 1992. The 6555th was deactivated when AFSC and Air Force Logistics Command merged to become Air Force Materiel Command on 1 July 1992.

Air Force managed programs
The Air Force's SATKA efforts involved experiments with space-based surveillance and tracking systems for locating and tracking ballistic missiles in various portions of their flight trajectories. Directed energy weapon experiments involved space- and ground-based lasers, and KEW programs supported research into ground- and space-based kinetic kill vehicles and electromagnetic launchers. Some of the programs were managed by Space Division's program offices in Los Angeles, but others were managed by Air Force laboratories or product divisions elsewhere in the United States. The Air Force's SDI budget for Fiscal Year 1985 was $845 million, but it more than doubled (to approximately $1.9 billion) in Fiscal 1986.

STS proposal
The Space Task Group, led by Vice President Spiro T. Agnew, presented the STS proposal to the President in the form of three alternatives on 15 September 1969. Each alternative included a manned reusable shuttle as part of a balanced program of space launch vehicles, and the Group recommended the Space Shuttle for a wide variety of DOD and NASA missions. As one of the many victims of budget cutbacks in the late 1960s, the Air Force's Manned Orbiting Laboratory (MOL) program had just been cancelled. Since the Air Force would not be able to field a manned space program of its own, the AFSC Commander (General James Ferguson) concluded that the use of the Space Shuttle for manned military space missions was essential. General Ferguson confirmed this point in a letter to the Air Force Chief of Staff in late October 1969. He also suggested the Air Force look into: 1) the kinds of military missions the Shuttle would support, 2) the extent of Air Force involvement in the program's management and 3) the scope of Air Force/ NASA agreements needed to ensure the Shuttle met military requirements.

Interim Upper Stage IUS
NASA expected to replace the Interim Upper Stage with NASA's Space Tug, but the Space Tug never materialized. On 16 December 1977, Air Force Assistant Secretary Dr. John J. Martin directed the Air Force to redesignate the IUS as the "Inertial Upper Stage."

higher energy orbits
The Air Force's interest in the IUS was based on the belief that more than half of all future DOD spacecraft would operate from high energy orbits. Since the Shuttle was confined to low-Earth

orbits, high-orbiting DOD payloads would require a Shuttle-compatible version of the IUS. The IUS would also be designed to operate with unmanned vehicles (e.g., Titan IIIC).

serious deficiencies
The problems included cracks in the small solid rocket motor nozzles, defective exit cones, and bad propellant.

Shuttle facilities
The responsibility for designing Shuttle launch and support facilities at Vandenberg was divided among the Air Force, the Navy and the U.S. Army Corps of Engineers. The Air Force and Navy hired contractors to design the projects assigned to them, but the Corps of Engineers did some of the design work itself. The Corps of Engineers and the Naval Facilities Command contracted out the actual construction, but Air Force Systems Command controlled the funds for the entire project and approved any major changes. The Space and Missile Systems Organization monitored the construction effort and obtained equipment for the facilities.

Challenger disaster
By the end of 1985, SLC-6's first Shuttle launch was scheduled for July 1986. Since the pad still lacked a hydrogen disposal system in January 1986, that launch date was questionable. On 28 January 1986, the Shuttle Challenger was engulfed in a fireball after its external tank exploded 73 seconds after lift-off from the Kennedy Space Center's Pad 39B. The Challenger disaster and its aftermath eliminated any possibility of a west coast launch after January 1986. In effect, added safety features made the Shuttle too heavy to fly missions from Vandenberg.

minimum caretaker status
The difference between operational caretaker status and minimum caretaker status related to the level of effort expended to preserve a site for future operations. Left in operational caretaker status, SLC-6 could have been readied for launch operations in two years. In minimum caretaker condition, the facility would require at least four years of concerted effort to prepare it for launch operations.

technical direction of Vandenberg's SCOUT launches to NASA.
NASA already supervised SCOUT missions launched from Wallops Island and the San Marcos platform, including military launches.

DELTA
Only the first two stages of the DELTA used liquid fuel (i.e., RP-1 for the first stage and Aerozine-50 for the second stage). Liquid oxygen was the oxidizer in the first stage, and the second stage used nitrogen tetroxide as its oxidizer. The second stage was capable of multiple starts to adjust the third stage and spacecraft's course. The third stage was a Thiokol solid rocket rated at approximately 14,000 pounds of thrust.

DELTA launch vehicle
In the 1960s, the DELTA was somewhat shorter. Its first stage was a 70-foot-long THOR rated at 172,000 pounds of thrust. From the 1970s onward, DELTAs was equipped with THOR first stages that were approximately 74 feet long. All THOR stages were eight feet in diameter. The DELTA's second stage was approximately 20 feet long and five feet nine inches in diameter. Its Aerojet ITIP engine generated more than 9,000 pounds of thrust. Seated atop the first stage, the second stage carried a miniskirt assembly (eight feet in diameter and eleven inches high) attached 42 inches from its top. The miniskirt attached to an interstage barrel that extended upward from the first stage. The interstage barrel was eight feet in diameter, and it gave the DELTA its "straight eight" (unbroken eight-foot diameter) profile. The payload fairing topped off the vehicle, giving it a height of 112 to 116 feet.

NASA
NASA supervised all DELTA missions launched from Space Launch Complex 2 (West) at Vandenberg and Complex 17 at the Cape during the 1970s and most of the 1980s. As we noted earlier, the Delta program and Complex 17 were transferred to the Air Force toward the end of 1988.

The first of those payloads
The first mission involved a Fleet Satellite Communications (FLTSATCOM) satellite, and it was accomplished under a contract between SAMSO and NASA to launch the spacecraft on a NASA ATLAS/CENTAUR vehicle. Though NASA and its contractors were responsible for the preparation and launch of the vehicle, the FLTSATCOM spacecraft was received, processed, checked out, stored and tested by the Air Force and its contractors. The launch on February 9th was successful, and the spacecraft became an integral part of the U.S. Navy's FLTSATCOM system and the Air Force's Satellite Communications System (AFSATCOM). The FLTSATCOM system provided instant, secure and reliable communications between the President of the United States and his military commanders as well as naval aircraft, ships, submarines and ground stations around the world. Seven more FLTSATCOM satellites were boosted into space from Launch Complex 36 over the next 11 years. (See Chapter III for details of those missions.)

family of heavy spacelifters
Though some may want to include the TITAN IIIE as an Air Force launch vehicle, the "E" was not, strictly speaking, an Air Force booster: it was supported by the Air Force and its contractors, but it was dedicated to NASA missions. In a joint-agency effort, NASA and the Air Force launched TITAN IIIEs from Complex 41 on a VIKING simulator mission and a HELIOS solar mission in 1974, two VIKING missions to Mars in 1975, another HELIOS mission in 1976 and two VOYAGER missions to the outer planets in 1977. The TITAN IIIE consisted of a standard core, two solid rocket motors and a CENTAUR upper stage.

standard core
The standard core consisted of two stages. Both stages were 10 feet in diameter, but the first stage was 71 feet long and the second stage was 37 feet long. Both stages burned a 50/50 mixture of

hydrazine and unsymmetrical dimethylhydrazine with nitrogen tetroxide as the oxidizer. In fueled condition, the standard core weighed 342,000 pounds, and it produced between 470,000 and 520,000 pounds of thrust when it was ignited at an altitude of 28 nautical miles (i.e., when boosted off the launch pad in TITAN IIIC or TITAN IIID configuration). The TITAN IIIB's stretched core was 68 inches longer than the standard core, but its power was somewhat less-this was due to the fact that the TITAN IIIB's core was ignited at lift-off rather than at altitude.

TITAN IIICs were launched only from Launch Complex 40
Complex 41 supported TITAN IIIC missions in the late 1960s, but it was devoted to TITAN IIIE space missions in the 1970s before its deactivation in 1977.

TITAN IIIC's and TITAN IIID
The solid rocket motors for the TITAN IIICs and Ds were provided by the Chemical Systems Division of the United Technologies Corporation. Aerojet Liquid Rocket Company provided the liquid rocket engines, and McDonnell Douglas provided the payload fairings for the TITAN IIICs. Delco Electronics manufactured the guidance sets, and the instrumentation systems came from SCI Systems. Actron provided the command destruct receivers. All those subsystems were delivered to Martin Marietta for integration with Martin's own TITAN airframes.

TITAN 34D would replace the TITAN IIIC and TITAN IIID
The TITAN IIIB program was unaffected by the decision to develop the TITAN 34D. TITAN IIIB operations continued at Space Launch Complex 4 (West) through February 1987. TITAN II space launches continued from that site in 1988.

orbiter's main engines
Space Shuttle orbiters were equipped with three main engines rated at approximately 390,000 pounds of thrust each at sea level. They all fired at lift-off, providing power along with two 149-foot-long solid rocket boosters rated at 2,650,000 pounds of thrust each. In its launch configuration, the orbiter was attached to a 154-foot-long external tank, which, in turn, was attached to the two solid rocket boosters. The assembled vehicle was approximately 184 feet tall.

Solid Motor Assembly Building (SMAB)
Space Division was responsible for developing the facilities that would be used to assemble and check out the IUS and mate it with its payloads. The first IUS processing operations would be carried out in the east end of the SMAB, but Shuttle payloads would be mated with IUSs in the west end of the SMAB later on.

three different problems
The first of those problems involved about 50 square feet of insulation that debonded from Columbia's external tank during a fuel-loading test in January 1981. Tank repairs pushed the launch back to 5 April 1981. A further delay of about five days was precipitated by a short labor strike against Boeing by machinists and aerospace workers. The third and final delay was caused by a computer anomaly which forced a launch scrub about 20 minutes before Columbia was scheduled to lift off on April 10th.

"lessons learned" reports

The 6555th's report on the subject, "82-1 Ground Processing Lessons Learned Summary," (dated November 1982) observed that Shuttle/DOD missions were manpower intensive and that a cadre of Defense Department officials (in addition to the 6555th's standard test team) would be needed to support a plethora of NASA meetings (e.g., scheduling meetings, procedure reviews, pretest meetings, communications, security etc.). A fundamental difference between Shuttle and unmanned launch operations was also noted under "General Observations" in the report: whereas the Test Group and KSC's Expendable Launch Director looked forward to a successful satellite deployment, NASA's overriding concern during a Shuttle mission was the successful launch and recovery of the orbiter. The payload, of necessity, took a backseat to the safety of the orbiter and its crew. The report went on to note that "the spacecraft community just doesn't have the clout it is accustomed to."

Columbia's STS-5 flight

Columbia was commanded by Vance Brand and piloted by Robert F. Overmyer on this mission. The mission specialists were Dr. William B. Lenoir and Dr. Joseph P. Allen. Columbia was launched from Pad 39A at 1219:00 Greenwich Mean Time on 12 November 1982, and the orbiter made a hard-surface runway landing at Edwards Air Force Base on 16 November 1982.

first extravehicular activity (i.e., spacewalk)

Dr. Lenoir and Dr. Allen were scheduled to perform the first spacewalk during STS-5, but a spacesuit malfunction forced the spacewalk's cancellation. Mission specialists Donald H. Peterson and Dr. F. Story Musgrave performed the first Shuttle spacewalk on STS-6.

Challenger's second mission (STS-7)

The satellites were designated ANIK C-2 (sponsored by TELESAT CANADA) and PALAPA B-1 (sponsored by Indonesia). The orbiter commander for the seven-day mission was Navy Captain Robert L. Crippen, and Navy Captain Frederick C. Hauck served as Challenger's pilot. The mission specialists for STS-7 were Dr. Sally K. Ride, Air Force Lt. Colonel John M. Fabian and Dr. Norman Thagard.

Challenger's third mission

One of the primary objectives of Challenger's third mission (STS-8) was communications link testing with the TDRS-A satellite deployed on STS-6. The tests were conducted successfully during days one through five of STS-8.

That flight (STS-8)

Challenger was commanded on STS-8 by Navy Captain Richard H. Truly. Commander Daniel C. Brandenstein piloted the orbiter. The mission specialists were Lt. Commander Dale A. Gardner, Air Force Colonel Guion S. Bluford, and Dr. William E. Thornton. Challenger touched down on Edwards' Runway 22 at 0740:43 Greenwich Mean Time on 5 September 1983.

Space Lab
Space Lab was a shared venture by the ten nations represented in the European Space Agency (i.e., France, Belgium, the Netherlands, Germany, Denmark, the United Kingdom, Spain, Italy, Switzerland and Austria). The Space Lab laboratory system consisted of a pressurized core, a long or short module for experiments, and an unpressurized pallet. The pressurized core was connected to the orbiter's cabin to provide a shirtsleeve work environment for payload specialists. Once Columbia was in orbit, the six-man crew worked in two three-man shifts to provide around-the clock Space Lab operations during the STS-9 mission. Space Lab was designed to be used up to 50 times during its 10-year lifespan, and it was flown on several Shuttle missions after its debut on STS-9.

CELV's concept definition
On 6 January 1984, Space Division asked Martin Marietta and General Dynamics' Convair Division to develop concepts for the CELV based on the TITAN 34D (for Martin Marietta) and the ATLAS (for Convair). Each contractor was to spend four months and no more than $500,000 to define the new vehicle in terms of the necessary upgrades.

NASA protested the CELV action
NASA Administrator James M. Beggs believed with some justification that any withdrawal of military payloads from the Shuttle's launch manifest tended to undermine financial support for the Space Transportation System by lowering the flight rate (at least on paper) and raising the average cost of Shuttle missions. Beggs filed his protests with the Secretary of the Air Force and the Secretary of Defense in May 1984. NASA also protested the CELV action to Congress during the same period.

CENTAUR G
The Air Force and NASA both concluded that a more powerful upper stage was needed for two DOD/Shuttle missions in 1987 and NASA's Galileo and International Solar Polar missions. The new upper stage - called the CENTAUR G (military version) and the CENTAUR G' (NASA version) - would be adapted from the CENTAUR upper stage already in use on the ATLAS/CENTAUR launch vehicle. The CENTAUR G would be 19.5 feet long and 14.2 feet in diameter (versus the ATLAS/CENTAUR's 30 x 10-foot upper stage). Since NASA's interplanetary missions required greater endurance, NASA planned to make the CENTAUR G' 29.1 feet long to store the extra fuel and oxidizer needed for the Galileo's mission to Jupiter and the International Solar Polar mission. The new CENTAUR G would be twice as powerful as the IUS, and it would be able to place a 10,000-pound payload into geosynchronous orbit from the Shuttle's cargo bay or, as proposed by Martin Marietta, from the top of a TITAN 34D7. A joint NASA/DOD CENTAUR working group was already hard at work on CENTAUR G/G' spacecraft requirements and upper stage configurations when Martin Marietta proposed the TITAN 34D7.

TITAN IV
Like its TITAN IIIC and TITAN 34D ancestors, the TITAN IV would be configured around three

stages. State O consisted of two seven-segment solid rocket motors that were 10 feet in diameter and 112.9 feet long. The motors weighed about 695,000 pounds apiece, and they provided a combined thrust of approximately 3,000,000 pounds at lift-off. They burned UTP-300B solid composite propellant during the first 111 seconds of flight, remained on the vehicle until about ten seconds after Stage I ignited, and separated from the liquid core at an altitude of about 175,000 feet approximately 126.4 seconds into the flight. Stage I was 10 feet in diameter and 86.5 feet long. It fired for the first time about 116.6 seconds into the flight, and its liquid turbopump-fed engine generated about 547,000 pounds of thrust from a 50/50 mixture of hydrazine and unsymmetrical dimethylhydrazine (UDMH). In a typical east coast launch, Stage II ignited approximately 302.1 seconds into the flight at an altitude of about 468,700 feet. Stage I was jettisoned less than a second later, and Stage II continued to burn its own supply of nitrogen tetroxide and UDMH for about 222.8 seconds. Stage II was 32.7 feet long and ten feet in diameter, and it weighed about 86,000 pounds at the start of its burn. It provided approximately 106,000 pounds of thrust at shutdown about 532.9 seconds into the flight. At the time of separation, Stage II and its payload were about 1,000 nautical miles downrange at an altitude of 532,000 feet. The payload subsequently entered a parking orbit (e.g., 80 x 95 nautical miles).

converted TITAN II missile/space boosters
The Strategic Air Command's inventory of 55 inactivated TITAN II ICBMs was transferred to AFSC and converted into a supply of space boosters to launch relatively small payloads into space. The first TITAN II missile/space booster was rolled out on 3 August 1987, and the first TITAN II space booster was launched from Vandenberg's Space Launch Complex 4 (West) on 5 September 1988.

On 28 August 1985, a TITAN 34D launch failure
An Air Force Class A Space Mishap Investigation Board was established under Brigadier General Donald J. Kutyna, and it examined the TITAN 34D mishap from September 3rd through October 18th. In its final progress report (issued 28 October 1985), the Board concluded the failure occurred in the first stage of the TITAN's liquid core. No "absolute" cause could be determined, but the evidence suggested the vehicle experienced an oxidizer propellant system leak as well as a turbopump subassembly failure. Since the TITAN required both engine subassemblies to maintain controlled flight, the subassembly failure caused the vehicle to go out of control. The turbopump failed when its pinion gear broke down due to loss of gear cooling, lubrication or some kind of pressurization loss-the ultimate cause remained unknown.

TITAN 34D was launched from Vandenberg on 18 April 1986
TITAN 34D-9 exploded eight seconds after lifting off Space Launch Complex 4 (East) on April 18th. Upper sections of the vehicle's solid rockets and bare fuel showered the launch pad, causing severe damage to launch facilities nearby. In some instances, large steel fragments were blown 3000 feet from the point of impact. The explosion also created a toxic cloud that rose to an altitude of 8000 feet before it was blown out over the Pacific Ocean. The AFSC Inspector General's Office selected the ESMC Commander, Brigadier General (Selectee) Nathan J. Lindsay, to serve as the president for the Mishap Investigation Board. The Board issued its final progress report on 9 June

1986, and that report suggested a variety of potential causes, mostly related to solid propellant/ insulation debonding.

recovery program
The Air Force initiated an intensive recovery program for the TITAN 34D and its ground support systems. The recovery included in-depth hardware inspections (featuring x-ray and other non-destructive tests), additional engine instrumentation, rocket motor joint heaters and procedural improvements. As Air Force Secretary Edward C. Aldridge noted after a TITAN 34D launch at the Cape in November 1987, the launches on both coasts emphasized "our confidence in the TITAN launch system and its ability to launch critical national security payloads in support of America's space launch recovery program."

launch failures of 1986
In addition to the Challenger disaster in January and the TITAN 34D-9 mission failure in April 1986, a DELTA carrying the GOES-G weather satellite broke up about a minute and a half after lift-off from the Cape on 3 May 1986. (Range safety officers sent arm and destruct commands to the vehicle at T plus 90.1 seconds). The immediate cause of the flight failure was an early main engine cutoff, but the root of the failure needed to be determined before more DELTAs were allowed to fly. Since there were similarities in the ATLAS and DELTA main engineering electronics relay boxes/wiring harnesses, the subsequent DELTA-178 accident investigation delayed testing support for both types of vehicles for several months. The next DELTA mission (DELTA-180) was launched successfully on 5 September 1986, and the next ATLAS-CENTAUR mission (AC-66) was launched successfully on 5 December 1986.

contract for13 additional TITAN IVs
That contract was definitized in December 1987. In addition to accelerating the launch rate, Martin agreed to launch some TITAN IVs from Vandenberg and add a No Upper Stage (NUS) configuration to the CENTAUR and IUS configurations already on contract. The total value of the (now) 23-vehicle contract was $4,420,000,000. It should be noted that the government did not accept any TITAN IV launch vehicle until the booster was at least one inch off the pad (following launch). Space Division agreed to pay a $7 million incentive fee for each flight success, but Martin Marietta would be penalized $45 million for each flight failure.

MLVs were needed
The TITAN II launch vehicle did not have sufficient thrust to boost 1,850-pound and 2,100-pound Block II satellites into their required 10,898-nautical-mile-high orbits. The TITAN IV could have been used, but it was clearly too big to boost NAVSTAR satellites into orbit economically. The MLV was needed to fill the gap between the TITAN II and TITAN IV. Under the new launch plan, the first NAVSTAR Block II satellite would be launched on an MLV around January 1989 - two years later than its previously estimated first launch date aboard the Shuttle. If all went well, a constellation of 18 GPS satellites and three orbiting spares could be in orbit by January 1991.

DELTA II

The DELTA II's first stage produced at least 207,000 pounds of thrust, and six of the DELTA II's nine solid rockets added approximately 97,000 pound of thrust (each) at lift-off. The DELTA II's three remaining solid rockets fired after lift-off, and they produced between 110,200 and 112,300 pounds of average thrust (each) at altitude. The DELTA's second stage was equipped with an Aerojet AJ10-18K engine rated at approximately 9,645 pounds of thrust at altitude. The Model 6925's solid rockets featured steel casings like the older DELTAs, but the Model 7925's solid rockets were known as Graphite Epoxy Motors (GEMs) because of their lighter graphite-epoxy casings. The GEMs and steel-cased Castor IVAs weighed 28,657 pounds and 25,562 pounds respectively, and all of the GEM's additional weight was translated into fuel. The GEMs were longer than the steel-cased Castor IVAs (e.g., 401.6 inches versus 323.4 inches), and they produced about 16 percent more impulse power (and 25,000 pounds more maximum power) than the steel-cased Castor IVAs. The GEMs' higher performance accounted for the margin of power the Model 7925 needed to boost heavier Block II satellites into 10,898-nautical-mile-high orbits.

CENTAUR G' (upper stage) project

The Shuttle-configured CENTAUR G' effort was managed by a NASA/Air Force joint program office at NASA's Lewis Research Center in Cleveland, Ohio. NASA retained overall management for the project, but Space Division provided five officers and a deputy project manager.

TITAN IV contract with Martin Marietta

Martin Marietta subcontracted its CENTAUR G upper stages to General Dynamics Space Systems Division.

review of the CENTAUR's design

Space Division remained committed to the TITAN IV/CENTAUR concept following that review. As Space Systems Division, the agency increased its procurement to 15 CENTAURs in September 1990.

DELTA program

Though NASA sponsored its last DELTA payload processing operation at the Cape in 1989, military and civilian DELTA space launches continued under Air Force sponsorship following the transfer of Complex 17 and its support facilities back to the Air Force in September and October 1988. McDonnell Douglas launched its first commercial DELTA mission from Pad 17B on 27 August 1989. As mentioned earlier, the first DELTA II/NAVSTAR mission was launched from Pad 17A on 14 February 1989.

ATLAS/CENTAUR vehicle

The ATLAS II's fuel tank was nine feet longer than the ATLAS G's fuel tank, and the CENTAUR upper stage was three feet longer. The ATLAS II's uprated Rocketdyne MA-5A engine system (i.e., two boosters and one sustainer) produced approximately 484,000 pounds of thrust. The ATLAS G and ATLAS II used the same CENTAUR engines (e.g., Pratt & Whitney RL10A-3-3A engines), but the ATLAS IIA's CENTAUR would be equipped with two upgraded Pratt & Whitney RL10A-4 engines rated at 20,800 pounds of thrust each.

DSCS III spacecraft

Three DSCS III spacecraft had been orbited previously for developmental purposes to demonstrate the satellite's concepts, systems and interfaces with ground stations. The first two operational DSCS III payloads were launched on the Shuttle. The DSCS III launched on February 11th was pushed into orbit by the first production Integrated Apogee Boost Subsystem (IABS). The DSCS III was orbited to support worldwide communications between military command posts and forces in the field.

"return to flight"

The flight of Discovery from 29 September through 3 October 1988 heralded the resumption of Shuttle missions after the Challenger tragedy. Discovery's mission was the culmination of a 32-month long recovery effort, which amounted to a thorough reappraisal of all Shuttle systems. During that period, the Shuttle's solid rocket boosters were redesigned (with special attention given to the rocket motor joints and seals). Most of the three surviving orbiters' major systems and components were removed and sent back to their vendors for inspection, modification or refabrication. When the Shuttle fleet stood down in 1986, Discovery's processing operations were reoriented toward the recovery: the orbiter was moved back and forth between the Vertical Assembly Building (VAB) and the Orbiter Processing Facility (OPF) for various modifications during the summer of 1986, and Discovery's major systems were returned to their vendors for modification in the fall. Flight processing began in mid-September 1987, and Discovery's three main engines were installed at the Kennedy Space Center (KSC) in January 1988. The newly redesigned solid rocket booster segments started arriving at KSC on March 1st, and booster stacking operations were completed between the end of March and the end of May 1988. The boosters were mated to Discovery's external tank on June 10th, and Discovery was moved from the OPF to the VAB on June 21st so the orbiter could be mated to the rest of the vehicle. Discovery rolled out of the VAB on the 4th of July and began its trip to Pad 39B. The Flight Readiness Firing was conducted successfully on 10 August 1988, and an Orbital Maneuvering System leak was repaired on August 19th. While minor problems were noted during the flight in September and October 1988, the mission was a complete success.

Chapter One Endnotes

1. Perry, Robert L., Origins of the USAF Space Program 1945-1956, Space Systems Division History Office, 1961, pp. 9, 12, 20, 34.

2. AFETR History, 1964, Volume I, p.3; Perry, Origins, pp. 35, 41, 44, 53, 55, 56, 57; ARDC History, 1 July - 31 December 1959, "Foreword" and "Chronology;" AFSCF History Office, "AFSCF History Brief and Chronology, 1954 - 1981," p. 2; Ley, Willy, Rockets, Missiles and Men in Space, N.Y. Viking Press, 1968 Edition, p 362; 6550th ABG History Office, "Chronology of Atlantic Missile Range and Air Force Missile Test Center, 1938 - 1959," p. 115; General Directive Number 1, NASA, "Proclamation on Organization of the National Aeronautics and Space Administration," 25 September 1958; Executive Order Number 10783, 1 October 1958.

3. History of the Assistant Commander for Missile Tests, AFBMD, 1 June - 20 December 1959, "Organization and Mission;" 6555th Test Wing (Development) History, 21 December 1959 - 31 March 1960, DWTI Historical Section, "Introduction" and "Missile Test Activities;" 6555th Test Wing (Development) History, 1 April - 30 June 1960, DWTI Historical Section, "Organization and Mission" and "Missile Test Activities;" AFMTC History, 1 January - 30 June 1957, p. 110; AFMTC History, 1 January - 30 June 1959, pp. 169, 170; AFMTC History, 1 July - 31 December 1959, pp. 179, 180; 6555th Test Wing (Development) History, 1 July - 31 December 1960, DWTI Historical Section, "Missile Test Activities".

4. 6555th Test Wing (Development) History, 1 January - 30 June 1961, DWS Historical Section, "Introduction", DWT Historical Section, "Organization and Mission", DWZI Historical Section, "Introduction," "Physical Facilities" and "Personnel", DWZS Historical Section, "Organization and Functions" and DWZT Historical Section, "Introduction;" 6555th Test Wing (Development) History, 21 December 1959 - 31 March 1960, DWOS Historical Section, "Introduction" and "Mission;" 6555th Test Wing (Development) History, 1 April - 30 June 1960, DWOS Historical Section, "Introduction" and "Missile Test Activities" 6555th Aerospace Test Wing History, 1 July - 31 December 1961, DWZ Historical Section, "Introduction" and DWZT Historical Section, "Introduction".

5. NASA, "National Aeronautics and Space Administration Agena B Launch Vehicle Program Management Organization and Procedures," 14 February 1961; NASA and 6555th Aerospace Test Wing, "Memorandum of Agreement on Participation of the 6555th Test Wing (Development) in the Centaur R&D Flight Test Program," 18 April 1961.

6. NASA and 6555th Aerospace Test Wing, "Memorandum of Agreement on Participation of the 6555th Test Wing (Development) in the Centaur R&D Flight Test Program," 18 April 1961, pp. 1, 2, 3 and Addendum.

7. 6555th Aerospace Test Wing History, 1 January - 30 June 1962, DWZC Historical Section, "Activities;" 6555th Aerospace Test Wing History, 1 July - 31 December 1962, DWZC Historical Section, "Organization and Mission," "Personnel" and "Activities;" 6555th Aerospace Test Wing History, 1 January - 30 June 1963, DWTC Historical Section, "Personnel Losses" and DWZC Historical Section, "Organization and Mission," "Strength Resume" and "Activities;" 6555th Aerospace Test Wing History, 1 July - 31 December 1963, DWZC Historical Section, "Activities" 6555th Aerospace Test Wing History, 1 January - 30 June 1964, DWZC Historical Section, "Activities;" 6555th Aerospace Test Wing History, 1 July - 31 December 1964, DWZ Historical Section, "Activities;" 6555th Aerospace Test Wing History, 1 January - 30 June 1965, DWC Historical Section, "Division Test Activities;" 6555th Aerospace Test Wing History, 1 January - 30 June 1966, DWC Historical Section, "Division Test Activities;" 6555th Aerospace Test Wing History, 1 July - 31 December 1966, DWC Historical Section, "Division Test Activities;" Crespino, Janice E., ESMC/HO, "Launches From the Eastern Test Range 1950 - 1990," April 1991, p. 4; Marven R. Whipple, "Atlantic Missile Range/Eastern Test Range Index of Missile Launchings," July 1963 - June 1964, p. 2; 6555th Aerospace Test Wing History, 1 January - 30 June 1967, DWC Historical Section, ("Mission, Objectives and Organization"); 6555th Aerospace Test Wing History, 1 July - 31 December 1967, DWC Historical Section, "Mission, Objectives and Organization;" 6555th Aerospace Test Wing History, 1 January - 30 June 1968, DWC Historical Section, "Division Test Activities;" 6555th Aerospace Test Wing History, 1 July - 31 December 1969, DWC Historical Section, "Strength Resume;" 6555th Aerospace Test Group History, 1 January -30 June 1970, ATLAS Systems Division Historical Section, "ATLAS Systems Division Test Activities".

8. 6555th Aerospace Test Wing History, 1 July - 31 December 1961, DWZS Historical Section, "Missile Test Activities;" 6555th Aerospace Test Wing History, 1 January - 30 June 1962, DWZS Historical Section, "Personnel" and "Missile Activities;" 6555th Aerospace Test Wing History, 1 July - 31 December 1963, DWZS Historical Section, "Test Program;" 6555th Aerospace Test Wing History, 1 January - 30 June 1964, DWZS Historical Section, "Test Program" and "Problem Areas;" 6555th Aerospace Test Wing History, 1 January - 30 June 1965, DWS Historical Section, "Missile Test Activities" and "Personnel Assignments;" 6555th Aerospace Test Wing History, 1 July - 31 December 1965, "Organization".

9. 6555th Aerospace Test Wing History, 1 July - 31 December 1961, DWZT Historical Section, "Introduction;" 6555th Aerospace Test Wing History, 1 January - 30 June 1962, DWZT Historical Section, "Introduction," "Physical Facilities" and Missile Test Activities;" 6555th Aerospace Test Wing History, 1 July -31 December 1962, DWZB Historical Section, "Introduction" and Program 624A Activities" and DWZT Historical Section, "Organizational Structure" and "Missile Test Activities;" Handbook, Martin Company, "USAF TITAN III Standard Space Launch System," 3rd Edition, undated, pp. A-2, A-3; 6555th Aerospace Test Wing History, 1 January - 30 June 1963,

DWZB Historical Section, "Introduction" and "Physical Facilities", DWZG Historical Section, "Introduction" and DWZT Historical Section, "Introduction" and "Missile Test Activities".

10. Whipple, "Index, July 1961- Jun 1962," p. 10; Whipple, "Index, July 1962- June 1963," pp. 5, 6; 6555th Aerospace Test Wing History, 1 January-30 June 1964, DWZT Historical Section, "Missile Test Activities;" 6555th Aerospace Test Wing History, 1 July - 31 December 1964, DWZ Historical Section, "Program Activities;" 6555th Aerospace Test Wing History, 1 January- 30 June 1965, DWT Historical Section, "Introduction" and "Activities;" Whipple, "Index, July 1963 - June 1964," p. 29; Whipple, "Index, July 1964 - June 1965," p. 31; USAF and NASA, "Agreement Between USAF and NASA for Transition of the Delta Space Launch Vehicle Program," signed 1 July 1988; Letter, Mr. James D. Phillips, to ESMC/DER, "Transfer of Real Property (Delta Program)," 18 August 1988; 6550th ABG/DER, "Weekly Activity Report," 4 October 1988.

11. 6555th Aerospace Test Wing History, 1 January - 30 June 1963, DWF Historical Section, "Facilities Activity" and DWZB Historical Section, "Organization and Mission" and "Program 624 Activities;" 6555th Aerospace Test Wing History, 1 July - 31 December 1963, DWZB Historical Section, "Introduction" and "Organizational Structure;" 6555th Aerospace Test Wing History, 1 January - 30 June 1964, DWZB Historical Section, "Introduction" and "Physical Facilities;" Whipple, "Index, July 1964 - June 1965," pp. 27, 29; Whipple, "Index, July 1965 -June 1966," p. 23; 6555th Aerospace Test Wing History, 1 July - 31 December 1964, DWB Historical Section, "

12. Special Order G-34, HQ AFSC, 30 March 1970; Message, HQ AFSC to HQ SAMSO, "Establishment of SAMTEC," 120019Z March 1970; Article, "6555th Test Wing Renamed in SAMSO Reorganization," The Missileer, 10 April 1970; Article, "Efficiency Dictates Test Center's Formation," Astronews, 17 April 1970; Headquarters Space Division History Office, "Space and Missile Systems Organization: A Chronology, 1954 - 1979," p. 12; Headquarters AFSC History Office, "Organizational Charts of the Air Force Systems Command, 1950 to Present," November 1987, pp. II-10 through II-14, II-16, II-19; 6555th Aerospace Test Wing History, 1 January - 30 June 1970, Chapter I, "Organization and Mission;" Whipple, "Index, July 1950 - 1960," p.20-2; Crespino, "Launches," introductory summary and pp. 27, 44; Study, HQ SAMTO, "Centroid Study Panel on Test Wings and Ranges," 21 November 1967; AFETR History, 1 January 1967 - 30 June 1968, Volume I, Part 2, pp. 319 through 322; AFETR History, Fiscal 1969, Volume I, Part 2, p. 271; AFETR History, FY 1970, Volume I, Part 2, p. 300; AFETR History, Fiscal Year 1971, Volume I, Part 2, p. 288.

13. AFETR History, FY 1973, Volume I, Part 1, p. 2; AFSC History, 1 July 1972 - 30 June 1973, Volume I, p. 3; AFSC History, 1 July 1973 - 30 June 1974, Volume I, p. 389; AFSC History, 1 January - 31 December 1977, Volume I, pp. 11, 12.

14. SAMTO/WSMC History, 1 October 1979 - 30 September 1980, pp. 3, 4, 5; A.F. Simpson Historical Research Center, "USAF Unit Lineage and Honors, Eastern Space and Missile Center," 27 November 1979.

15. SAMSO History, 1 January - 31 December 1977, Volume I, pp. 1, 2; SAMSO History, 1 January - 31 December 1978, Volume I, pp. 1, 2; SAMSO History, 1 January - 31 September 1979, Volume I, pp. 1, 2.

16. Space Division History, 1 October 1981 - 30 September 1982, Volume I, pp. 12, 13, 14; AFSPACECOM History, January - December 1990, pp. 1, 2; AFSC History, 1 October 1982 - 30 September 1983, Volume I, pp. 21, 22.

17. Space Division History, 1 October 1982 - 30 September 1983, Volume I, pp. 93, 98; Space Division History, 1 October 1984 - 30 September 1985, Volume I, pp. 160, 161; Space Division History, 1 October 1985 - 30 September 1986, Volume I, p. 149; Space Division History, 1 October 1986 - 30 September 1987, Volume I, pp. xliv, xlv; Space Division History, 1 October 1987 - 30 September 1988, Volume I, pp. 6, 183, 184; Space Systems Division History, October 1989 - September 1990, Volume I, pp. 365, 357; SAMTO/WSMC History, 1 October 1987 to 30 September 1988, Volume I, pp. 9, 10.

18. ESMC History, 1 October 1989 - 30 September 1990, Volume I, pp. 383, 384; 9th Space Division History, 1 October 1990 - 30 September 1991, p. 3; Headquarters AFSPACECOM Programming Plan 90-2, 15 August 1990, p. 3; 45 SPW History, 1 October 1990 - 31 December 1991, Volume I, pp. 39, 40; Article, "Second Atlas II launch activates new squadron," The Missileer, 10 July 1992; Item, "Group deactivates after merger," The Missileer, 10 July 1992; Interview, Mark C. Cleary with Mr. Jeffrey Geiger, 30th Space Wing Historian, 7 July 1993.

19. Space Division History, 1 October 1983 - 30 September 1984, Volume I, pp. 332, 333, 334 and 335; AFSC History, 1 October 1985 -30 September 1986, Volume I, pp. 236, 237, 238; Space Division History, October 1987-September 1988, Volume I, p. 446.

20. AFSC History, 1 July 1974 - 30 June 1975, Volume I, pp. 329, 331, 333, 334, 336; AFSC History, 1 January - 31 December 1977, Volume I, p. 477, 479, 484; AFSC History, 1 October 1981 - 30 September 1982, Volume I, p. 291.

21. AFSC History, 1 July 1974 - 30 June 1975, Volume I, pp. 338, 342, 345; AFSC History, 1 January - 31 December 1977, Volume I, p. 483, 486, 487, 488, 489; Space Division History, October 1987 -September 1988, Volume I, p. 156.

22. SAMSO History, 1 January - 31 December 1978, Volume I, pp. 100, 101; Space Division History, 1 October 1979 - 30 September 1980, Volume I, pp. 77, 78, 79, 80, 81, 82, 83; Space Division History, 1 October 1980 - 30 September 1981, Volume I, p. 105, 116, 117, 119, 120, 123; Space Division History, 1 October 1981 - 30 September 1982, Volume I, pp. 78, 81.

23. SAMSO History, 1 January - 31 December 1978, Volume I, pp. 110, 111, 112, 113, 114; Space Division, 1 October 1980 - 30 September 1981, Volume I, pp. 127, 128; SAMSO History, 1

January - 30 September 1979, Volume I, p. 87; Space Division History, 1 October 1981 - 30 September 1982, Volume I, p. 90.

24. ESMC History, 1 October 1984 - 30 September 1986, Volume I, pp. 342, 347; SAMTO/ WSMC History, 1 October 1986 - 30 September 1987, Volume I, pp. 148, 149, 150, 151; SAMTO/ WSMC History, 1 October 1987 - 30 September 1988, Volume I, pp. 139, 140, 141.

25. Fact Sheet, AFMTC Office of Information, "Scout Launch Vehicle Fact Sheet," 30 July 1963; SAMSO History, 1 January - 31 December 1978, Volume I, pp. 51, 54.

26. Fact Sheet, NASA, "Delta," December 1969; Fact Sheet, NASA, "Delta," January 1975; Press Kit, NASA/KSC, "Delta 182/Palapa-B2P Press Kit," 16 March 1987, p.3; Interview, Mark C. Cleary with Mr. Jeffrey Geiger, 30th Space Wing Historian, 14 July 1993; SAMSO History, 1 January - 31 December 1978, Volume I, pp. 51, 52.

27. 6555th Aerospace Test Group History, January - September 1978, p. 21; 6555th Aerospace Test Group History, 1 January - 30 June 1971, Atlas Systems Division Section, "Part IV;" AFETR History, Fiscal Year 1973, p. 355.

28. 6555th Aerospace Test Group History, January - September 1978, p. 21; Det 1 SAMTEC History, January - December 1978, Volume I, pp. 119, 120; ESMC History, 1 October 1979 - 30 September 1980, Volume I, p. 354; Crespino, "Launches," p. 6; ESMC History, 1 October 1989 - 30 September 1990, Volume I, pp. 108, 109, 302, 303; Fact Sheet, NASA, "Atlas/Centaur," March 1976; ESMC History, 1 October 1982 - 30 September 1984, Volume I, p. 203; ESMC History, 1 October 1986 -30 September 1987, Volume I, p. 358; ESMC History, 1 October 1988 -30 September 1989, Volume I, p. 354.

29. Handbook, The Martin Company, "USAF TITAN III Standard Space Launch System," Third Edition, undated, pp. A-1, A-2, A-3, A-4; SAMSO History, 1 January - 31 December 1978, Volume I, pp. 70, 71, 72; SAMSO History, 1 January - 31 December 1979, Volume I, pp. 53, 54; Cleary, Mark C., The 6555th: Missile and Space Launches Through 1970, 45th Space Wing History Office, 1992, p. 207; Gieger, Jeffrey, "Vandenberg AFB Launch Summary," 30th Space Wing History Office, 15 December 1992, pp. 58 through 95;

30. SAMSO History, 1 January - 31 December 1978, Volume I, pp. 77; SAMSO History, 1 January - 31 September 1979, Volume I, p. 67, 68; Space Division History, 1 October 1980 - 30 September 1981, Volume I, pp. 90, 92; ESMC History, 1 October 1982 - 30 September 1984, Volume I, p. 165; ESMC History, 1 October 1984 - 30 September 1985, Volume I, p. 223; ESMC History, 1 October 1988 - 30 September 1989, Volume I, pp. 344; Geiger, "Launch Summary," pp. 88 through 97.

31. SAMSO History, 1 January - 31 September 1979, Volume I, pp. 73, 74, 75, 76; Space Division

History, 1 October 1979 - 30 September 1980, Volume I, pp. 67, 68, 75, 76; Pan American World Airways, Inc. and RCA International Service Corporation, "Range Pretest Briefing, Space Shuttle OFT," April 1981, p. 3.

32. History of Space Division, 1 October 1981 - 30 September 1982, Volume I, pp. 108, 109; ESMC History, 1 October 1979 - 30 September 1981, Volume I, p. 96; ESMC History 1 October 1982 - 30 September 1984, Volume I, pp. 60, 61.

33. Space Division History, 1 October 1980 - 30 September 1981, Volume I, pp. 101, 103, 104; NASA, "STS-1 First Space Shuttle Mission Press Kit," April 1981, pp. 1, 2, 3, 47, 48, 50, 51; ESMC History, 1 October 1981 - 30 September 1982, Volume I, pp. 184, 187, 189, 190; 6555th ASTG, "82-1 Ground Processing Lessons Learned Summary," a/o November 1982, pp. T-2, O-1; Crespino, "Launches," p. 46.

34. Message, DDMS to AIG 7016, "Launch Schedule Status Report (L-60 Days)," 261400Z November 1982; ESMC History, 1 October 1982 - 30 September 1984, Volume I, pp. 179, 180, 181, 182, 183.

35. Message, DDMS to AIG 7016, "STS-7 Launch Schedule Status Report," 161810Z April 1983; Message, DDMS to AIG 7016, "STS-8 Launch Schedule Status Report," 061500Z July 1983; ESMC History, 1 October 1982 - 30 September 1984, Volume I, pp. 184, 185, 186, 197, 188, 189.

36. ESMC History, 1 October 1982 - 30 September 1984, Volume I, pp. 189 192, 193, 194; Pan Am World Services and RCA International Corporation, "STS-9 Space Shuttle Range Pretest Briefing," 28 November 1983, pp. 2, 5; Message, DDMS to AIG 7016, "STS-9 Launch Schedule Status Report," 061900Z September 1983.

37. ESMC History, 1 October 1982 - 30 September 1984, Volume I, pp. 194, 195.

38. Space Division History, 1 October 1983 - 30 September 1984, Volume I, pp. 80, 89, 90; Space Division History, 1 October 1984 - 30 September 1985, Volume I, pp. 99, 100, 101.

39. Space Division History, 1 October 1983 - 30 September 1984, Volume I, pp. 90, 91, 92, 94; Space Division History, 1 October 1984 - 30 September 1985, Volume I, pp. 99, 100; Space Division History, October 1985 - September 1986, Volume I, pp. 124, 125; ESMC/45 SPW History, 1 October 1990 - 30 September 1991, Volume I, pp. 316, 317.

40. Geiger, "Launch Summary," a/o 15 December 1992, pp. 94, 95, 97; Space Division History, October 1986 - September 1987, Volume I, p. xlii; ESMC History, 1 October 1984 - 30 September 1986, Volume I, pp. 43, 223; Space Division History, October 1985 - September 1986, Volume I, pp. 67, 86, 87, 91, 93, 94; R&D Associates, "Fuel Impact Explosion Study," Volume I, p. 1-1; ESMC History, 1 October 1987 - 30 September 1988, Volume I, p. 286.

41. ESMC History, 1 October 1984 - 30 September 1986, Volume I, pp. 282, 283, 284; ESMC History, 1 October 1986 - 30 September 1987, Volume I, pp.359, 360, 362; SAMTO/WSMC History, 1 October 1986 - 30 September 1987, Volume I, pp. 101, 102.

42. Space Division History, October 1985 - September 1986, Volume I, pp. 67, 68, 69, 70, 253, 254; Space Division History, October 1986 - September 1987, Volume I, pp. 53, 54; Space Division History, October 1987 - September 1988, Volume I, pp. 98, 99.

43. Space Division History, October 1986 - September 1987, Volume I, pp. 55, 56; Space Division History, October 1987 - September 1988, Volume I, pp. 66, 71, 72; ESMC History, 1 October 1990 - 31 December 1991, Volume I, pp. 155, 156, 332, 334; ESMC History, 1 October 1988 - 30 September 1989, Volume I, pp. 359, 360, 364.

44. Space Division History, October 1984 - September 1985, Volume I, p. 126; Space Division History, October 1985 - September 1986, Volume I, pp. 125, 126, 127.

45. Space Division History, October 1985 - September 1986, Volume I, pp. 127 and 128; Space Division History, October 1987 - September 1988, Volume I, p. 97; Space Systems Division History, October 1989 - September 1990, Volume I, p. 103.

46. ESMC History, 1 October 1988 - 30 September 1989, Volume I, pp. 161, 162, 170, 358; ESMC History, 1 October 1989 - 30 September 1990, Volume I, pp. 108, 109, 110; Space Division History, October 1987 - September 1988, Volume I, pp. 78, 80, 81; ESMC and 45 SPW History, 1 October 1990 - 31 December 1991, Volume I, p. 331; 45 SPW History, 1 January - 31 December 1992, Volume I, pp. 219, 221, 222, 225.

47. ESMC History, 1 October 1987 - 30 September 1988, Volume I, p. 276; Crespino, "Launches," pp. 46, 47; ESMC History, 1 October 1982 - 30 September 1984, Volume I, p. 198; ESMC History, 1 October 1984 - 30 September 1986, Volume I, pp. 238, 241, 245, 258, 261; ESMC History, 1 October 1987 - 30 September 1988, Volume I, pp. 276, 277, 278, 286; ESMC History, 1 October 1988 - 30 September 1989, Volume I, pp. 334, 342; ESMC History, 1 October 1989 - 30 September 1990, Volume I, pp. 285, 286, 287; ESMC and 45 SPW History, 1 October 1990 - 31 December 1991, Volume I, pp. xi, 298, 305, 306, 313; 45 SPW History, 1 January - 31 December 1992, Volume I, p. ix.

48. AFSC History, 1 October 1982 - 30 September 1983, Volume I, p. 325.

The Cape, Chapter 2, Section 1

TITAN and Shuttle Military Space Operations

6555th Aerospace Test Group Responsibilities

The history of military space operations at the Cape is bedecked with great engineering feats and gilded with spectacular space flights. The forces behind those achievements were often channeled to a good end by the diligent efforts of the 6555th Aerospace Test Group and its work with other agencies (e.g., U. S. Army Corps of Engineers, launch vehicle and spacecraft contractors, range contractors and higher headquarters). It is only fitting that we continue our examination of military space operations with an overview of the 6555th Test Group's organization, duties and responsibilities (and those of its successors) as set forth in Air Force documents in the early 1970s, late 70s, mid-80s and early 1990s. This introduction applies equally to chapters II and III, since both medium and heavy launch vehicle systems will be addressed.

At the beginning of 1971, the 6555th Aerospace Test Group consisted of a commander's office under Colonel Davis P. Parrish and three divisions (e.g., Support, ATLAS Systems and TITAN III Systems). The Group's overall mission was to provide field test management and launch support for AFSC and other agencies at the Cape. Its responsibilities included: 1) representing the Space and Missile Systems Organization (SAMSO) and the 6595th Aerospace Test Wing at the Cape in the areas of technical test direction and program control, 2) integrating Air Force, other government agency and contractor efforts in support of program field test management, prelaunch and launch support, 3) determining the test and/or launch readiness of launch vehicles and payloads and 4) providing liaison between the 6595th Aerospace Test Wing and the Air Force Eastern Test Range (AFETR) organization. Under the Test Group's concept of launch operations, Air Force launch operations engineers were placed "on the scene" to work closely with booster, payload and range support contractors. They observed individual and combined systems tests, and they helped contractors resolve problems and meet test objectives successfully under the pressure of time constraints. Based on their familiarity with those systems, the Test Group's "blue suit" engineers evaluated trend data and performed failure analysis in concert with the contractors. Their principal duties were those of field test engineers.[1]

Though the Test Group's launch operations revolved around the ATLAS and TITAN III systems divisions in the early 1970s, the Group established its Space Transportation System (STS) Division on 1 July 1974 to ensure the Defense Department's Shuttle requirements were factored into future Shuttle operations at the Kennedy Space Center (KSC). Under the direction of Lt. Colonel Morgan W. Sanborn, the five-member STS Division set up shop in KSC's Headquarters Building and began working STS ground support system issues with the NASA Shuttle Project Manager (Dr. Robert Gray) and his staff. The Division's responsibilities included development of the STS Ground Operations Plan and definition of the facilities, equipment, security and safety systems that would be needed.[2]

Figure 36: Colonel Davis P. Parrish

On 1 November 1975, the Test Group reorganized its ATLAS and TITAN III launch vehicle agencies under a new division, the Space Launch Vehicle Systems Division. On the same date, the ATLAS Satellite Launch Systems Branch and the TITAN III Space Satellite Systems Launch Operations Branch were consolidated under the newly-created Satellite Systems Division. The changes were directed by the 6595th Aerospace Test Wing Commander to combine booster operations under one division chief and payload operations under another division chief. In the same vein, the IUS Operations Branch was placed under the Space Launch Vehicle Systems Division when that branch was formed on 1 July 1977. Following the final ATLAS/AGENA launch on 6 April 1978, the Space Launch Vehicle Systems Division and the Satellite Systems Division shifted their respective attentions from ATLAS/AGENA operations on Complex 13 to ATLAS/CENTAUR boosters and payloads designated for Defense Department missions on Complex 36.[3]

When the 6555th Aerospace Test Group was transferred from the 6595th Aerospace Test Wing to the Eastern Space and Missile Center on 1 October 1979, the Group's three divisions were left intact. The Test Group created its Programs/Analysis Division around April 1981, but that division dealt with budget and facility planning matters, and it reinforced rather than diminished the basic missions of the other three divisions. Later on, following the first Shuttle missions, the Air Force saw a need to streamline Shuttle/DOD payload operations by simplifying "interfaces" between NASA, the Air Force and various contractors. The STS Division and the Satellite Systems Division were consolidated to form the Spacecraft Division on 1 November 1983. Until June 1988, the 6555th based its organization on the Space Launch Vehicle Systems Division, the Spacecraft Division and the Programs/Analysis Division.[4]

Under the KSC/6555th ASTG Joint Operations Plan for DOD Missions (dated 7 January 1985), the Spacecraft Division directed ground processing of Defense Department payloads and determined the technical readiness of spacecraft, ground support equipment and facilities. Its Air Force test controllers managed spacecraft hardware testing in the Shuttle Payload Integration Facility (SPIF), and the Division provided the Air Force Test Director for Space Shuttle missions involving Defense Department payloads. The Space Launch Vehicle Systems Division exercised field test management and control over all TITAN 34D, IUS, CENTAUR, and TITAN IV vehicles and upper stages associated with military missions launched from KSC and the Cape. The Programs/Analysis Division processed documents related to the Test Group's responsibilities and its plans for the future. In the summer of

1987, there was talk of creating a "DELTA Division" to handle the Test Group's new DELTA II launch vehicle program. On 1 June 1988, 20 manpower authorizations were transferred from the Space Launch Vehicle Systems Division to form the initial cadre for the Medium Launch Vehicle Division. (The Spacecraft Division transferred three of its manpower slots to the new division as well.) Under the 6555th Test Group's charter, the Medium Launch Vehicle Division became the focal point for all launch site activities related to medium launch vehicles. The new division would provide engineering direction for booster, upper stage and payload activities, and it would certify the vehicle *and* the payload for launch.[5]

The next important shift in the 6555th Test Group's organization occurred as a result of ESMC's transfer from Air Force Systems Command (AFSC) to Air Force Space Command (AFSPACECOM) on 1 October 1990. On that date, 175 out of 241 personnel were transferred "on paper" from the 6555th to the 1st Space Launch Squadron, the ATLAS II and TITAN IV Combined Test Forces (CTFs), "Payload Operations" and "Ops Resource Management." The Test Group's remaining personnel remained attached to the 6555th, but their numbers dwindled to approximately 25 military members and 11 civilians by the end of December 1991. Colonel Michael R. Spence assumed command of the 6555th Aerospace Test Group on 2 October 1990, and he was given an additional position on the ESMC staff as Deputy for Launch Operations. Under this new "dual-hatted" position, Spence supervised the resources formerly assigned to the 6555th Aerospace Test Group.[6]

Figure 37: Colonel Michael R. Spence

Following the inactivation of ESMC and the creation of the 45th Space Wing on 12 November 1991, the Cape's various military launch and spacecraft organizations were locked into place. The 1st Space Launch Squadron was reassigned from ESMC to the 45th Operations Group. The ATLAS II and TITAN IV CTFs were placed under the 45th Operations Support Squadron until such time as they could become fully operational squadrons in their own right. (The TITAN IV CTF became merely the "TITAN CTF" under the 45th Operations Squadron, and the ATLAS II CTF became the ATLAS Division.) The Ops Resource Management Office became the 45th Operations Support Squadron's Launch Operations Support agency.

The Payload Operations Office became the 45th Spacecraft Operations Squadron. Colonel Spence succeeded Colonel James N. Posey as 45th Operations Group Commander on 31 January 1992, and Lt. Colonel William H. Barnett became the 6555th Test Group's "Acting Commander" in addition to his duties as Director of the TITAN CTF. Lt. Colonel Barnett assumed command of the 6555th in his own right on 25 March 1992. Air Force Systems Command and Air Force Logistics Command (AFLC) were consolidated into Air Force Materiel Command (AFMC) on 1 July 1992, and the 6555th Aerospace Test

Group was deactivated. The Test Group's liaison function between AFSPACECOM and Space Systems Division was assumed by Detachment 8 of the Space and Missile Systems Center. Following the second successful launch of a military ATLAS II/CENTAUR mission on 2 July 1992, the ATLAS Division was activated as the 3rd Space Launch Squadron.[7]

Figure 38: 45th Operations Group Emblem

TITAN and Shuttle Military Space Operations

Launch Squadron Supervision of Military Space Operations in the 1990s

So much for the basic organization of military space operations at the Cape. At this point we may ask, "*how* were military space operations supervised at the Cape?" For a constructive answer to that question, we must rely on launch and spacecraft agency training aids after military space operations were transferred to AFSPACECOM. (As a result of the organizational continuity mentioned earlier, it is reasonable to assume that the space launch squadrons' operating instructions and the TITAN CTF and Spacecraft Operations Squadron self-study guides were derived from earlier practices and formalized.) The training aids give us a brief but detailed look into the world of military space operations.[8]

The 1st and 3rd Space Launch Squadron commanders had overall responsibility for launch operations in their respective squadrons, but they relied on a highly trained and educated team of officers and non-commissioned officers to conduct day-to-day operations at the Cape. Each squadron's Space Launch Operations Controllers (SLOCs) acted as the Air Force's on-scene representatives during booster processing operations, and they ensured safety and security standards were maintained and proper procedures were followed to process launch vehicles and analyze any problems that surfaced during the course of operations. The SLOCs were authorized to stop operations any time procedures, safety, or security standards were violated. In the event of an accident, the SLOCs took action to minimize injuries and equipment damage. They also preserved evidence of the mishap until relieved by proper authority.[9]

Systems Engineers (SEs) performed a variety of technical roles in both squadrons. They reviewed electronic and mechanical hardware modifications to hydraulic, pneumatic and propellant systems, structures and ordnance. They monitored individual contractor actions as they occurred, and they reviewed and approved highly complicated procedures ahead of time. Their duties included participation in engineering "walkdowns" (led by McDonnell Douglas and General Dynamics launch operations managers) to detect vehicle damage and improper hardware installation at the launch pads. During the walkdowns, SEs reported any discrepancies they detected immediately, and they followed up with the appropriate company's engineering managers to resolve any problems that could not be corrected on the spot. Individual SEs were also tasked to serve as Vehicle Engineers (VEs) for each launch vehicle. Vehicle Engineers monitored the status of all test procedures, vehicle problems, site concerns and ground equipment issues. The senior VE briefed launch vehicle status at the Launch Readiness Review and tasked SEs to resolve technical problems. He also provided the engineering "go/no-go" recommendation to the Launch Controller (LC) on launch day.[10]

Countdown operations were among the most crucial activities in the entire launch preparation process, and the 1st Space Launch Squadron's officials played important roles on the Air Force/McDonnell Douglas launch team responsible for those countdown sequences. The Squadron's portion of the team consisted of a Launch Director (LD), a Launch Controller (LC), a Launch Operations Manager (LOM), SLOCs, a Facility Operations Manager (FOM), a Booster Countdown Controller (BCC), SEs, a Facility Anomaly Chief (FAC) and an Anomaly Team Chief. (See figures 39 and 40 for team duty locations on launch day.) The Launch Director reported directly to the Mission Director (MD) and commanded the launch crew through its prelaunch, launch, postlaunch and launch abort/scrub activities. The Launch Controller, Anomaly Team Chief and Launch Weather Officer reported to the Launch Director. The Launch Director authorized continuation of the countdown following built-in holds in the countdown, and he also made the final launch vehicle go/no-go recommendation to the Mission Director before launch.[11]

The Launch Controller controlled operations at Space Launch Complex 17 during countdown preparations, terminal countdown and the period the site was secured following countdown. The LC received status reports from McDonnell Douglas' Launch Conductor and other members of the Squadron's launch crew. Given the length of a typical DELTA II countdown, there were actually two LCs for each DELTA II mission: the first shift LC was on duty from the initial launch day crew brief through completion of initial checklist items. The second shift LC relieved the first shift LC and remained on duty until the launch complex was secured from countdown. On either shift, the LC was responsible for making sure the team was ready to move ahead through key milestones in the checklist. The LCs approved deviations in procedure, and they coordinated the team's reaction to local weather conditions and flight and ground system problems.[12]

The Launch Operations Manager provided contact between the LC in the blockhouse and SLOCs on the launch pad. In essence, that duty involved receiving and reporting pad and vehicle status to the LC before terminal countdown. The LOM verified the launch complex's readiness for launch operations. This included the complex's voice communications capability, and its closed circuit television and camera coverage of the launch.[13]

The Facility Operations Manager was responsible for the launch complex's readiness (e.g., towers, fire control systems, electrical power, generators, fuel tanks, water systems and other support items.) The Launch Base Services (LBS) contractor assigned to the complex reported to the FOM, and the FOM controlled that contractor's operations on the pad, except for pad safety. The FOM verified the facility's readiness to support the Mobile Service Tower's rollback and the site's closeout before terminal countdown.[14]

For countdown operations, the Mission Vehicle Engineer (VE) was selected to be the launch team's

Booster Countdown Controller (BCC) because he possessed the most comprehensive knowledge of the technical aspects of the launch vehicle's processing history (except for third stage and payload systems). The BCC controlled the Squadron's engineering team and ensured problems detected by the SEs on that team were brought to the attention of the Anomaly Team Chief and the Launch Controller. The SEs supervised the following operations during the countdown:[15]

1. First stage propellant loading.

2. Second stage propellant and pneumatic tank pressurization.

3. Launch vehicle and ground support equipment electrical systems (assigned to two SEs).

4. Third stage electrical systems.

5. Solid rocket motor, first and second stage pneumatic thruster pressure and launch vehicle air-conditioning (assigned to two SEs).

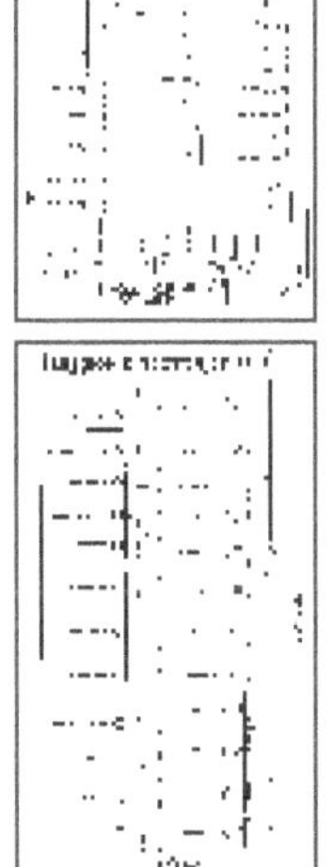

Figure 39: Launch Vehicle Data Center

Figure 40: Mission Director Center

The Facility Anomaly Chief served as the Squadron's focal point for launch problems involving equipment and systems maintained by the LBS contractor (i.e., Johnson Controls World Services, Inc.). The FAC received countdown status from LBS contractor personnel and the Facility Operations Manager, and he reported technical concerns to the Anomaly Team Chief. The FAC was responsible for Space Launch Complex 17's electrical power, searchlights, pad deluge systems and air-conditioning and ventilation systems.[16]

The Squadron Chief Engineer served as the Anomaly Team Chief (ATC) for each launch vehicle. As such, the ATC was the single point of contact for vehicle problems. He evaluated the information he received from the FAC and reported problem solutions to the Launch Director. The ATC also reported on wind factors affecting the launch vehicle's trajectory, its structural load margins and guidance control margins.[17]

The 3rd Space Launch Squadron's launch team at Complex 36 was organized in the same manner as the 1st Space Launch Squadron's launch team on Complex 17. The Launch Director was in charge of ATLAS II countdown activities (e.g., prelaunch, postlaunch and launch abort/scrub activities), and he authorized continuation of the countdown through the built-in holds. He provided a final go/no-go recommendation to the Mission Director. The Launch Controller controlled countdown activities and reported to the Launch Director. Like their counterparts in the 1st Space Launch Squadron, the ATLAS II launch team's SLOCs conducted tower walkdowns, monitored the Mobile Service Tower's rollback, supervised launch pad closeout tasks, oversaw launch preparations by the contractor and reported status back to the Launch Controller. In similar fashion, the 3rd Space Launch Squadron Vehicle Engineer served as the Booster Countdown Controller for the mission.[18]

The TITAN CTF did not achieve full squadron status until 14 April 1994, but it began operating as the Launch Vehicle Directorate under the 45th Operations Group in the summer of 1992. The Directorate's two principal branches were the Operations Branch and the Vehicle Engineering Branch. The Operations Branch managed TITAN IV field operations, and it controlled TITAN IV ground hardware. The Vehicle Engineering Branch provided technical direction for flight hardware (e.g., TITAN IV core stages, solid rocket motors and payload fairings), and its CENTAUR and IUS engineering sections supervised CENTAUR and IUS upper stage operations.[19]

Roles and responsibilities in the Launch Vehicle Directorate were organized along the lines of duties in the 1st and 3rd Space Launch Squadrons. The TITAN Launch Director (LD) had overall responsibility for planning and conducting launch operations. He received information from the Launch Controller (LC) and the upper stage Vehicle Manager (VM), and he made the final go/no-go recommendation to the Mission Director (MD). The TITAN Launch Controller (LC) exercised operational control over the entire launch vehicle, including the upper stage. The CENTAUR Vehicle Manager (CVM) supervised CENTAUR operations. Through the CENTAUR Test Team, the CVM provided the final upper stage launch readiness recommendation to the LC. In similar fashion, the IUS Vehicle Manager (IVM) supervised IUS operations and served as the IUS team chief. The IVM was responsible for booster/IUS and payload/IUS activities, and he reported IUS status during the launch countdown.[20]

Figure 41: Blockhouse 17 Pre-Launch Alert 1991

The Launch Vehicle Operations Controller (LVOC) supervised day-to-day TITAN booster activities including schedules, testing, hardware removal/replacement and discrepancy reporting. As the launch vehicle team chief, the LVOC reported engineering status to the LC during major tests and launch countdowns. TITAN SLOCs monitored TITAN operations to ensure proper test discipline, security and safety. Like their 1st and 3rd space launch squadron counterparts, the TITAN SLOCs could stop operations whenever they detected an unsafe or improper operation. They minimized injuries and equipment damage in the event of an accident, and they preserved mishap evidence until they were relieved by proper authority.[21]

The TITAN Vehicle Engineer (VE) was responsible for assessing the technical readiness of the TITAN IV vehicle and its associated ground equipment. The VE provided vehicle status updates to the upper stage vehicle manager (CVM or IVM), the Launch Controller and the Launch Director. Systems Engineers were assigned to oversee various launch vehicle systems, and they provided the VE with the specific information he needed to form his assessments and make his reports.[22]

Figure 42: Hangar AE Control Center February 1991

TITAN and Shuttle Military Space Operations

TITAN IV Launch Contractors and Eastern Range Support Contractors

At this point, a short summary of the TITAN IV's principal contractors and subcontractors is in order. (ATLAS and DELTA contractors will be presented in Chapter III.) The Martin Marietta Corporation was the prime contractor for the TITAN IV launch vehicle. The corporation built the TITAN core stages, fabricated the vehicle's support equipment, activated the TITAN launch sites, integrated the launch vehicle with its payload and conducted the launch operation. Apart from all of those items and services, Martin relied heavily on subcontractors for the following systems:[23]

1. Liquid Rocket Engines (Aerojet Technical Systems).

2. Solid Rocket Motors (United Technologies, CSD).

3. Guidance System (DELCO System Operations).

4. CENTAUR Upper Stage (General Dynamics Space Systems).

5. Payload Fairing (McDonnell Douglas).

6. Instrumentation (SCI Systems).

7. Command/Control Receivers (Cincinnati Electric Corporation).

Figure 43: TITAN III Core Vehicle Fuel and Oxidizer Tanks

In addition to those booster-related contractors, Boeing Aerospace Operations (BAO) was the major contractor for the IUS. The IUS was manufactured and assembled in Seattle, Washington and shipped to the Cape for launch processing. The company used hangars E, F and H for its operations as well as facilities in the Integrate Transfer Launch (ITL) Area of the Cape. Unlike the subcontractors mentioned earlier, BAO was a prime contractor on a par with Martin Marietta. Martin and Boeing were considered associate contractors serving the same program.[24]

Figure 44: Aerojet TITAN IIIC Liquid Rocket Engine

Two other major contractors need to be mentioned before we move on. The first was Computer Sciences Raytheon (CSR). It was a joint venture partnership between Computer Sciences Corporation (CSC) and the Raytheon Company. Under the supervision of the 45th Space Wing Commander and his agents, CSR operated and maintained the Eastern Range for all range users in accordance with the Air Force's Range Technical Services (RTS) contract. Under that contract, CSR acquired and reduced launch vehicle and spacecraft flight data for the Launch Vehicle Directorate and other users. The partnership operated, modified, maintained and managed range systems (e.g., optics, radars, telemetry, command/destruct generators, and communications) located at the Cape, Merritt Island, Jonathan Dickinson Missile Tracking Annex, other Florida annexes, on the USNS *Redstone* and at downrange stations on the islands of Antigua and Ascension.[25]

Figure 45: TITAN Solid Rocket Motor Stacking

The other contractor was Johnson Controls World Services. As the Launch Base Services (LBS) contractor at the Cape, Johnson Controls provided launch pad safety, security, fire protection, Skid Strip operations, utilities and heavy equipment services for the Launch Vehicle Directorate. Johnson Controls was also responsible for facility planning, maintenance and repair services (e.g., plumbing, electrical work, carpentry, painting, sand-blasting, masonry, etc.). Like CSR, Johnson Controls was supervised by the 45th Space Wing Commander and his agents.[26]

TITAN and Shuttle Military Space Operations

Quality Assurance and Payload Processing Agencies

Quality assurance was an essential part of successful military space operations at the Cape. Under ESMC and the 45th Space Wing, military officials remained sensitive to quality assurance issues. They demanded high quality services and hardware from their contractors, who maintained their own staffs of quality assurance people. The 6555th Aerospace Test Group and its successors monitored military space operations with quality in mind, but the principal responsibility for quality assurance was carried out by civil servants in another part of ESMC and the 45th Space Wing. In the early 1980s, ESMC assigned the quality assurance role to the Quality Assurance Division under the Center's Directorate of Contracting and Support. The Division became a directorate following an ESMC reorganization in April 1986, but its duties continued to revolve around the following highly specialized programs:[27]

Figure 46: IUS Processing

1. Space Shuttle Solid Rocket Booster recovery and refurbishment.

2. NASA and Defense Department payloads.

3. Inertial Upper Stages.

4. TITAN Launch Vehicles.

By the end of the 1980s, 49 civil servants were assigned to the Directorate's Solid Rocket Booster Division and Quality Engineering Division to supervise Shuttle solid rocket booster recovery and refurbishment activities. Forty-six civilians also operated out of the Aerospace Systems Division and the Space Launch Systems Division to monitor IUS and TITAN quality assurance. Payloads were inspected by 32 civilians in the Space Transportation Systems Division, which was renamed the MLV and Payloads Division on 12 June 1990. Following the 45th Space Wing's activation in November 1991, the

Directorate of Quality was reorganized as a "three-letter" office (LGZ) under the 45th Logistics Support Squadron. Its divisions became four-letter offices with no reduction in the scope of their activities. They continued to provide quality assurance and complement inspection efforts in the space launch squadrons, the Launch Vehicle Directorate, and the 45th Spacecraft Operations Squadron.[28]

In our brief review of responsibilities, we have saved the 45th Spacecraft Operations Squadron for last because of its pivotal role in military payload activities. The mission of the 45th Spacecraft Operations Squadron (45 SPOS) was to serve as an executive agent for the Space and Missile Systems Center and process Defense Department spacecraft for flights on ATLAS, DELTA, TITAN and Shuttle vehicles. (The Squadron also processed other payloads as requested by other customers.) On paper, the Squadron had separate flights for ATLAS, DELTA, TITAN and Shuttle payload operations, but only the DELTA Payload Operations Flight had dedicated officers and non-commissioned officers (NCOs)-at least through 1992. The rest of the flights were supplied by a pool of officers and NCOs who were funneled into the separate flights as needed. As of November 1992, 58 people (including six officers from Air Force Materiel Command) were assigned to the 45 SPOS.[29]

On the contractor side of the "payload" house, the McDonnell Douglas Space Systems Company (MDSSC) provided approximately 150 people under the Launch Operations Support Contract (LOSC) to support operations at the Spacecraft Processing and Integration Facility (a.k.a., the Shuttle Payload Integration Facility) and work at the Operations Support Center (OSC). Rockwell International Corporation provided 80 people to process NAVSTAR Global Positioning System satellites launched on DELTA II boosters. General Electric Astro Space employed 100 people to process Defense Satellite Communications System (DSCS) payloads launched on the ATLAS II, and Johnson Controls dedicated 28 people to facility maintenance and other ground support functions. All together, more than 400 military people, civil servants and contractors supported spacecraft operations under the 45th Spacecraft Operations Squadron's taskings.[30]

The Air Force payload operations team was the linchpin for 45 SPOS spacecraft activities. The team was formed approximately one to two years before a spacecraft arrived at the Cape, and the team consisted of a Payload Operations Director (POD), a Field Program Manager and Deputy (FPM and DFPM), a Lead Non-Commissioned Officer (LNCO), Operations Controllers (OCs) and a Spacecraft Countdown Controller (SCC). The Payload Operations Director was responsible for all spacecraft launch base operations involving a particular booster class of payload (i.e., DELTA, ATLAS, TITAN or Shuttle). The POD managed and controlled spacecraft testing, fueling, prelaunch and launch activities for the Mission Director. He also selected the Field Program Manager, who would keep the team updated on the spacecraft's status.[31]

The Field Program Manager exercised control over the contractor's field processing and launch activities. Those workloads stepped up dramatically about six to twelve months before a launch, and the FPM had to ensure that the launch base was ready to receive flight hardware at the proper time. Test procedures were reviewed and approved, facilities were configured, and the payload operations team was briefed. After the spacecraft arrived at the Cape (about six months before launch), the FPM chaired daily spacecraft scheduling meetings with the POD, the Deputy FPM, the Payload Support Contractor, the Launch Systems Integration Contractor, the Launch Vehicle Contractor, the SPO, the Air Force Launch Controller, the Aerospace Corporation's representative and (often) a representative from Air Force Quality Assurance (LGZ). Based on the meetings, the FPM approved the spacecraft schedule and made sure operations were performed according to that schedule. The Deputy FPM assumed the FPM's duties during the latter's absence and attended to spacecraft security badging and test procedures.[32]

The Lead Non-Commissioned Officer was responsible for scheduling all range support activities for the spacecraft program. He received most of those requirements at the spacecraft scheduling meetings, but his experience and oversight were required to identify critical requirements that might have been missed in the documents presented at those meetings. The LNCO coordinated the spacecraft's arrival with the Operations Support Center and other support agencies approximately three days before the spacecraft arrived at the Cape. He was present during most major spacecraft operations, and, through his contacts, he was able to get support for missed services-especially those identified as "show-stoppers." The LNCO did not operate independently. He coordinated his requests with the Payload Support Contract Test Conductor. The work was scheduled through the Operations Support Center.[33]

The 45th Spacecraft Operations Squadron's Operations Controllers were responsible for controlling and supervising individual field operations (e.g., fueling, ordnance installation, solid motor buildup, spacecraft lifts, etc.). Like the SLOCs in the launch squadrons, the spacecraft OCs had the authority to stop processes whenever safety or security standards were violated. The OCs ensured that clean room (contamination control) standards were maintained in the processing areas. In the event of a spacecraft anomaly or processing accident, the OC stopped the operation and alerted the FPM.[34]

This brings us to the final member of the payload operations team-the Spacecraft Countdown Controller (SCC). When qualified, the FPM served as the SCC on launch day, but regardless of who served, the SCC was responsible for the performance of pre-countdown and countdown procedures on major tests and launch operations. The SCC ensured spacecraft readiness and participated in trouble-shooting processes during the countdown. He reported to the Payload Operations Director and the Launch Controller, and he conferred with the Payload Support Contract Test Conductor and the SPO on issues that might warrant a launch hold. At predetermined points in the countdown, the SCC provided go/no-go recommendations, including the final go/no-go recommendation before lift-off.[35]

TITAN and Shuttle Military Space Operations

TITAN IIIC Military Space Missions after 1970

We are now ready to go back to the beginning of 1971 to look at TITAN IIIC military space missions in detail. To set the scene briefly, Lt. Colonel Julius R. Conti was Chief of the 6555th Test Group's TITAN III Systems Division. The Division consisted of 23 officers, 42 enlisted people and 13 civilians, and it monitored operations in the Integrate-Transfer-Launch (ITL) Area, the Engineering and Analysis Building, Guidance Lab #2, the Satellite Assembly Building (SAB), the Air Force Spin Test Facility, Hangar L and Missile Assembly Buildings I and II. In the early 1970s, Launch Complex 41 was being modified for TITAN IIIE/CENTAUR operations to support NASA's VIKING (Mars Explorer) program. Consequently, TITAN IIIC launches were limited to Complex 40.

Figure 47: Colonel Robert D. Woodward

Figure 48: Colonel Arthur W. Banister

Figure 49: ITL Area Facilities

The principal contractors who carried out TITAN IIIC operations at the Cape were Martin Marietta, United Technologies, DELCO Electronics and the Aerojet Liquid Rocket Company. (The Aerospace Corporation also provided supervision and assistance.) Taken together, those companies' employees outnumbered the TITAN III System Division's people by about four to one. A Defense Support Program (DSP) mission had just been launched from Complex 40 on 6 November 1970, but the payload failed to achieve proper orbit. (The spacecraft's operational potential was reduced as a result.) The next TITAN IIIC vehicle and its DSP payload were assembled and checked out for a launch on 5 May 1971. The launch on May 5th was successful, and the payload was placed in a synchronous earth orbit as planned. That flight marked the 16th TITAN IIIC mission in the 6555th Test Group's history.[36]

On 2 November 1971, the Air Force and its contractors launched the first two Phase II Defense Satellite Communications Program (DSCP) satellites into near synchronous equatorial orbits from Complex 40. That TITAN IIIC mission was successful, and it marked the first in a series of classified flights destined to replace Phase I DSCP satellites that had been launched from the Cape between 16 June 1966 and 14 June 1968. On 1 March 1972, a TITAN IIIC carrying a 1,800-pound DSP satellite was launched successfully from Complex 40. Eight days later, a TITAN IIIC core vehicle (C-24) arrived via C-5A aircraft, and it was erected at the VIB on 16 March 1972. The Acceptance Combined Systems Test (CST) for that vehicle was completed on May 22nd, and the vehicle was accepted by the Air Force on 2 June 1972. Another core vehicle (C-26) arrived at the Skid Strip on June 30th. It was erected at the VIB on 5 July 1972, and it was accepted on 19 January 1973. Core Vehicle C-27 arrived on 3 May 1973, and it was erected in the VIB by 21 May 1973. In the meantime, preparations for the next TITAN IIIC launch went ahead: Core Vehicle C-24 was mated to its solid rocket motors and a DSP payload, and it was launched successfully on 12 June 1973. All booster and instrumentation systems performed well on the flight, and the desired 19,316 x 19,322-nautical-mile synchronous orbit was achieved.[37]

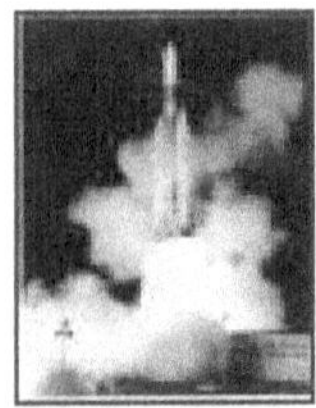

Figure 50: TITAN IIIC launch 5 May 1971

Figure 51: Vertical Integration Building

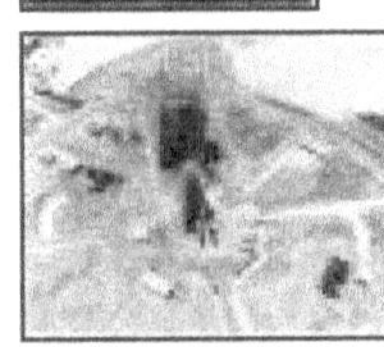

Figure 52: TITAN IIIC Complex 1965

Figure 53: TITAN IIIC launch 13 December 1973

Under Lt. Colonel Edwin W. Brenner, the TITAN III Systems Division supervised the preparation and lift-off of Launch Vehicle C-26 on a classified mission on 13 December 1973. The mission had been "scrubbed" (i.e., cancelled during countdown) on December 11th due to an airborne instrumentation support problem, but the countdown on December 13th went smoothly. The vehicle was launched from

Complex 40 without incident, and all booster and instrumentation systems performed well. The vehicle carried its payload into orbit successfully and injected two Phase II DSCP satellites into the proper orbits. The highly successful mission also marked the first use of the Universal Space Guidance System (USGS).[38]

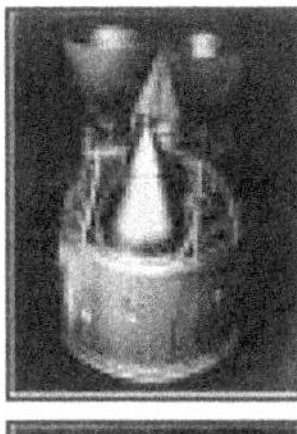

Figure 54: TITAN III TRANSTAGE

Figure 55: Artist Conception of Transtage ignition following Stage II separation

Figure 56: Cape Industrial Area

Launch Vehicle C-27 was processed by an Air Force/contractor team, but it was used for NASA's Applications Technology Satellite mission (ATS-F) that lifted off Complex 40 on 30 May 1974. The Air Force's support for that flight met all the primary and secondary mission objectives, and the vehicle achieved a satisfactory 19,334 x 19,302-nautical-mile final orbit. The next two TITAN IIIC vehicles (e.g., C-25 and C-28) had arrived at the Cape by that time, and both vehicles were being prepared for military missions. Launch Vehicle C-25 was scheduled to boost two Defense Satellite Communications System (formerly known as DSCP) spacecraft into orbit on 13 May 1975, but the countdown was delayed almost a week due to lightning strikes in the vicinity of the launch pad on May 9th. (Further prelaunch testing after the lightning strikes verified the integrity of the booster and its payload.) Another postponement moved the countdown to May 20th, but Launch Vehicle C-25 lifted off Complex 40 at 13:03 Greenwich Mean Time on that date. Though the initial stages of the flight went well, a guidance system power supply failure made it impossible to fire the transtage after the vehicle and payload entered parking orbit. Consequently, the DSCS satellites were marooned until their highly elliptical orbits decayed, and they reentered Earth's atmosphere a few weeks later.[39]

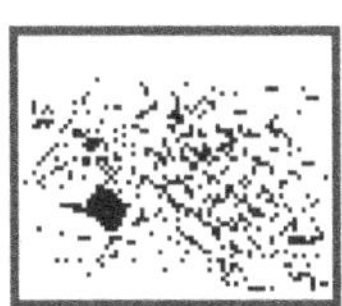

Figure 57: Map Cape Industrial Area

While Core Vehicle C-28 remained in storage at the VIB, a more recent arrival (Core Vehicle C-29) was prepared for a classified military mission. Core Vehicle C-29 was moved from the VIB to the SMAB on 20 October 1975. Following the mate with its solid rocket motors, the vehicle was moved out to Complex 40 on October 27th. Unlike the previous TITAN IIIC mission on May 20th, C-29's flight on 14 December 1975 was flawless, and the mission was a success. The next TITAN IIIC (C-30) had been accepted on 23 October 1975, and it was moved to the SMAB for solid rocket mating on December 5th. Following the C-29 mission on the 14th, C-30 was moved out to Complex 40 on 22 December 1975.[40]

Figure 58: Colonel John C. Bricker

Figure 59: Lt. Colonel Walter S. Yager

As operations continued to move ahead, there were significant changes in the Test Group's command positions and organization. On 24 June 1975, Colonel John C. Bricker assumed command of the 6555th Aerospace Test Group. This action followed Colonel Robert D. Woodward's retirement on June 1st. Lt. Colonel Edwin W. Brenner continued as Chief of the TITAN III Systems Division for most of 1975, but the Division was merged with the ATLAS/AGENA Launch Operations Branch and the ATLAS Mission Management Branch to form the Space Launch Vehicle Systems Division on 1 November 1975. Lt. Colonel Warren G. Green had been Chief of the ATLAS Systems Division before the reorganization. Green now became the Chief of the Space Launch Vehicle Systems Division. Lt. Colonel Walter S. Yager succeeded Lt. Colonel Green as Division Chief in the early part of 1977, and Lt. Colonel George L. Rosenhauer replaced Colonel John C. Bricker as Commander of the 6555th Aerospace Test Group on 8 June 1977.[41]

Figure 60: Brigadier General Don M. Hartung officiates at 6555th ASTG Change of Command Ceremony on 8 June 1977. Lt. Colonel Rosenhauer (near right) and Colonel Bricker (far right) standby.

Under Lt. Colonel Green, the Space Launch Vehicle Systems Division supervised the countdown and lift-off of Launch Vehicle C-30 on 14 March 1976. Two Lincoln Experimental Satellites (LES 8 and LES 9) were placed in synchronous orbit during that successful space flight, and C-30 also boosted the

SOLRAD 11A and 11Bspacecraft into 65,000-mile-high circular orbits to monitor solar conditions. Elsewhere in the ITL Area, Core Vehicle C-28 was taken out of storage and returned to its transporter in the VIB on 29 January 1976. The vehicle required extensive retesting after its extended stay in storage, but C-28's retest CST was completed successfully on 9 April 1976. The Core Vehicle was moved to the SMAB on April 15th, and the vehicle and solids were mated. The vehicle was moved out to Complex 40 on April 27th, and the Launch CST was completed successfully on 10 June 1976. With a classified payload onboard, Launch Vehicle C-28 lifted off Complex 40 at 2300:01 Eastern Daylight Time on 25 June 1976. The launch was a success.[42]

Figure 61: Flight Model, Lincoln Experimental Satellite

Figure 62: Martin Marietta's Denver Division Plant

No other TITAN IIICs were launched during the remaining months of 1976, but three vehicles were in various stages of preparation for upcoming military missions. The first of those vehicles was Core Vehicle C-31. It arrived at the Cape on 30 June 1976. The first two stages of the vehicle were erected on Transporter #3 in Cell #4 of the VIB on July 27th, and C-31 was moved to Cell #2 on 11 August 1976. The transtage was added to the core vehicle on August 12th, and an Acceptance CST was completed successfully on October 28th. The vehicle's hardware acceptance was completed on 9 November 1976. With no immediate launch date in view, Core Vehicle C-31 was moved to Cell #3 in the VIB where it was placed in storage on 6 December 1976. Preparations for the next vehicle (C-23) were well underway by that time. (Core Vehicle C-23 had been accepted at Martin Marietta's plant in Denver in May 1971, but its latest CST was completed in the VIB on 3 August 1976.) The vehicle was moved from the VIB to the SMAB for solid rocket mating on 14 December 1976. Following that operation, Launch Vehicle C-23 was moved out to Complex 40 on 21 December 1976. The third vehicle was Core Vehicle C-32. It arrived at the Cape on December 1st, and it was erected on Transporter #3 on 7 December 1976.[43]

Due to the classified nature of its payload, not much can be said about Launch Vehicle C-23's mission, but some of the events leading up to the launch are releasable. An eight-day readiness countdown was started for the vehicle on 25 January 1977 for a scheduled launch date of February 2nd. Unfortunately, the Launch CST was aborted on the first day of that countdown due to a problem with Stage I's hydraulic actuator. Complications arising from that incident pushed the launch to 5 February 1977, and the launch countdown did not get underway until February 4th. More bad luck: an automatic hold occurred at T minus 3 seconds when Stage I's Destruct Safe and Arm mechanism failed to arm. The countdown was recycled 24 hours, and it was picked up again on February 5th. Following a smooth and uneventful countdown on the night of February 5th, Launch Vehicle C-23 lifted off Complex 40 at 0100:00 Eastern Standard Time on 6 February 1977. According to an unclassified pamphlet published by the 6555th Aerospace Test Group in 1980, "all aspects of the mission occurred as planned".[44]

Figure 63: Defense Satellite Communications System (DSCS) II Satellite

Figure 64: SMAB (in foreground) and Complexes 40 (top right) and 41 (top left)

Core Vehicle C-32 was prepared for the next TITAN IIIC mission to be launched from Complex 40. Following a successful Acceptance CST on 24 February 1977, the core vehicle was accepted on 3 March 1977. On March 9th, the core vehicle was moved to the SMAB where it was mated to its solid rocket motors. Launch Vehicle C-32 was moved out to Complex 40 on March 16th, and it was mated to its twin DSCS (communications satellite) payload on 22 April 1977. Following a successful payload CST on April 25th, a seven-day readiness countdown began on May 3rd. Though the Launch CST went well, a Stage I engine ordnance connector was broken on the fifth day of the readiness countdown, and two days had to be added to the schedule to repair the connector. That problem aside, the launch countdown went smoothly on May 12th, and Launch Vehicle C-32 lifted off the pad at 1027:01 Eastern Daylight Time. The mission went well, and both DSCS Phase II satellites were injected into near synchronous equatorial orbits at an altitude of 19,334 nautical miles.[45]

Three more TITAN IIIC core vehicles arrived at the Cape in 1977. The first of them Core Vehicle C-33 arrived on March 16th. It was erected on Transporter #1 by March 25th, and its Acceptance CST was completed successfully on May 25th. Following hardware acceptance on 7 June 1977, Core Vehicle C-33 was moved to a storage stand in Cell #3 of the VIB. The next TITAN IIIC core vehicle (C-35) arrived on June 9th, and it was erected on Transporter #1 in Cell #2 by June 14th. An Acceptance CST was performed on C-35 on 1 August 1977, and the vehicle was moved to provide room for the third core vehicle. That vehicle (C-34) arrived at the Cape on October 5th. It was erected on Transporter #3 in Cell #2 by October 19th, and its Acceptance CST was performed successfully on 12 December 1977.[46]

Figure 65: TITAN IIIC being transported from the SMAB to Complex 40

Three TITAN IIIC missions were launched from Complex 40 during 1978. The first of those flights involved Core Vehicle C-35, which was mated to its solid rocket motors in the SMAB in late January 1978. Launch Vehicle C-35 was moved out to the launch pad on January 31st. On February 8th, the vehicle's twin DSCS payload arrived, and Remote Vehicle Checkout Facility (RVCF) tests were completed on the two communications satellites in that mission package on February 16th. Following booster/payload mating on March 5th, Launch Vehicle C-35 was fueled on March 7th, and its payload fairing was installed on March 11th. The TITAN IIIC lifted off Complex 40 on 25 March 1978 at 1309:00 Eastern Standard Time. Sadly, all efforts came to naught: a failure in the launch vehicle's Stage II hydraulic system occurred approximately eight minutes after lift-off, and Stage II's engine shut down prematurely; range safety officers sent arm and destruct commands to the TITAN IIIC, and the vehicle was destroyed.[47]

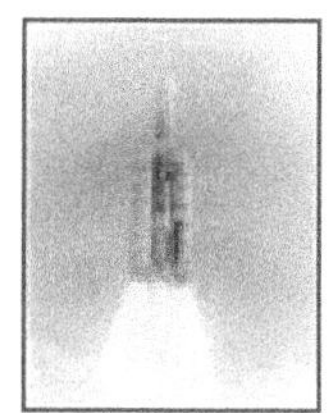

Figure 66: Launch Vehicle C-35 shortly after lift-off 25 March 1978

The second TITAN IIIC mission in 1978 involved Launch Vehicle C-33 and a classified payload. Due to the classified nature of that mission, we may only note that the vehicle was launched successfully on 10 June 1978 and "all aspects of the mission occurred as planned." The last TITAN IIIC mission of 1978 included two DSCS satellites and Launch Vehicle C-36. Core Vehicle C-36 arrived at the Cape on 14 June 1978, and that vehicle's subsystem testing began on July 11th. The vehicle was accepted formally on September 19th, and its solid rocket motors were installed after C-36 was moved to the SMAB on September 21st. Launch Vehicle C-36 was moved out to Complex 40 on September 26th, but an unacceptable buildup of martyte on one of the vehicle's solid rocket motors required a short return trip to the SMAB on September 29th. The vehicle was moved out to Complex 40 once again around October 4th, and it was mated to its DSCS payload on November 1st. The Launch CST was performed successfully on December 5th, and the vehicle was launched successfully at 1940:00 Eastern Standard Time on 13 December 1978. Both DSCS satellites were placed in their proper orbits and "drifted" into position over the eastern and western Pacific Ocean.[48]

Three TITAN IIIC launch vehicles (C-31, C-34 and C-37) were launched from Complex 40 in 1979. They supported two classified missions and one DSCS communications satellite mission. The oldest of the three vehicles (C-31) was launched first, after its second trip to the launch pad on 19 March 1979. Due to classification restrictions, not much can be said about C-31's countdown or mission. The vehicle lifted off Complex 40 on 10 June 1979 at 0930:00 Eastern Daylight Time and "all aspects of the mission occurred as planned."[49]

Vehicle C-34 was the next TITAN IIIC launched from the Cape. (As we noted earlier, Core Vehicle C-34 arrived at the Cape on 5 October 1977, and its Acceptance CST was performed successfully on 12 December 1977.) Like C-31, C-34 supported a classified mission, and the vehicle did not have a direct route from the VIB and the SMAB to the launch pad and space: Core Vehicle C-34 was designated to support initial testing of the TITAN III program's new Programmable Aerospace Checkout Equipment (PACE). Under that program, C-34 was placed on Transporter #2 in Cell #1 of the VIB on 4 January 1978, and it supported PACE testing in the VIB from 4 February through 7 November 1978. The vehicle was then moved to the SMAB for solid rocket mating on 3 January 1979, but its initial trip to Complex 40 on January 9th merely supported more PACE testing at the launch pad. The vehicle was moved back to the SMAB for solid rocket demate on March 5th, and it went back to the VIB on 21 March 1979. Following an Aerojet special engine retest in Cell #2 on March 21st, the core vehicle was moved into Cell #4 for storage (on its transporter) on April 17th. It was moved to Cell #2 again on 13 June 1979 to replace some hydraulic tubing. Following a retest of core vehicle systems, C-34 was moved to the SMAB for a second solid rocket remate on July 9th. Launch Vehicle C-34 was moved out to Complex 40 on July 13th, and it lifted off the pad on 1 October 1979 at 0722:00 Eastern Daylight Time. All aspects of the mission occurred as planned.[50]

The last TITAN IIIC mission in 1979 involved Launch Vehicle C-37 and a twin-DSCS payload. The core vehicle arrived at the Cape on 27 September 1978, and it was erected on Transporter #2 in Cell #2 of the VIB by 10 October 1978. Its Acceptance CST was run on 12 July 1979, and the vehicle was moved to the SMAB for solid rocket mating on 5 October 1979. Launch Vehicle C-37 was moved out to Complex 40 on 12 October 1979. We may assume that the prelaunch preparations were less complicated than the tasks leading up to the two previous TITAN IIIC missions. The Launch CST was run on 12 November 1979. Regarding the launch itself, there was one unscheduled five-minute hold during C-37's launch countdown on November 20th. The countdown resumed without further incident, and the TITAN IIIC lifted off Complex 40 on the 20th at 2110:00 Eastern Standard Time. Both Phase II DSCS communications satellites were placed in their proper near synchronous orbits, and the mission was a complete success.[51]

No TITAN IIICs were launched in 1980, and only two TITAN IIICs (C-40 and C-39) lifted off Complex 40 in 1981. Launch Vehicle C-40 was processed first, after its core arrived at the Cape on 9 April 1980. Core Vehicle C-40 was erected on Transporter #1 in Cell #1 of the VIB by 6 May 1980, and the Acceptance CST was completed on 25 July 1980. Following its formal acceptance by the Air Force on August 8th, Core Vehicle C-40 was transferred to the storage stand in Cell #3. (Transporter #1 was used shortly thereafter to support Core Vehicle C-39's assembly and acceptance testing.) A Retest CST was run on C-40 on 9 December 1980, and the core vehicle was moved to the SMAB for solid rocket mating on December 15th. Launch Vehicle C-40 was moved out to Complex 40 before the end of 1980, and its Launch CST was completed in early March 1981. Details of its classified mission are not releasable, but Launch Vehicle C-40 lifted off Complex 40 on 16 March 1981 at 1424:00 Eastern Standard Time.[52]

Figure 67: Titan IIIC mission

20 November 1979

Core Vehicle C-39 arrived at the Cape on 10 September 1980, and its erection was completed on Transporter #1 in Cell #1 by 15 September 1980. The vehicle's Acceptance CST was completed on 30 October 1980, and the vehicle was transferred to Transporter #3 and Cell #2 for storage on 15 November 1980. The vehicle was transferred to Cell #1 for additional work on 28 January 1981, and a Retest CST was completed on May 26th. Core Vehicle C-39 was moved to the SMAB for solid rocket mating on 22 June 1981, and it was moved out to Complex 40 around the end of the month. The Launch CST was completed in October 1981. Launch Vehicle C-39 carried a classified payload, and it lifted off Complex 40 on 31 October 1981 at 0422:00 Eastern Standard Time.[53]

Figure 68: TITAN IIIC Core Vehicle Erection 1980

The last vehicle launched under the TITAN IIIC program was Launch Vehicle C-38. It arrived at the Cape on 24 October 1979, and it was the last of 36 TITAN IIICs launched from the Cape between 18 June 1965 and the evening of 6 March 1982. After two years of testing, storage and retesting, C-38 was launched on a classified mission on 6 March 1982 at 1425:00 Eastern Standard Time. The flight marked the end of an era at the Cape.[54]

TITAN and Shuttle Military Space Operations

TITAN 34D Military Space Operations and Facilities at the Cape

As the last TITAN IIIC thundered skyward, Martin Marietta and the Test Group were completing their second year of preparations for the TITAN 34D's first launch. The effort began in earnest when the first TITAN 34D core vehicle (D-01) arrived at the Cape in March 1980. Baseline CSTs were completed in September 1980, and, apart from a brief roundtrip ride to the SMAB on November 11th, the core vehicle remained in storage at the VIB until 18 May 1981. Preparations for subsystem testing extended from late March through early June, and the Primary Acceptance CST was accomplished toward the end of June 1981. The core vehicle was accepted in August 1981, and it was moved to the SMAB on 18 January 1982. The core and solids were mated, and the launch vehicle was moved to Complex 41 pending cleanup on Complex 40 after the TITAN IIIC launch on March 6th. Launch Vehicle D-01 was moved to Complex 40 on March 24th, and it was powered up and mated to the IUS Pathfinder Test Vehicle (PTV-C) on April 1st. The PTV-C was utilized in conjunction with the D-01 launch vehicle in a Baseline CST, four umbilical drop tests, a two-week-long series of electromagnetic compatibility tests and a launch readiness verification test. Following the initial tests, aft/forward fairings and a model spacecraft were installed to permit a payload fairing electromagnetic compatibility CST on June 22nd. The PVT-C was demated from the launch vehicle on July 22nd, and it was disassembled and returned to the contractor. In the meantime, the TITAN 34D's operational IUS (IUS-2) arrived at the Cape on 22 December 1981. It was taken to the SMAB, and its assembly was completed there on 8 June 1982. Though the IUS' preplanned acceptance testing was completed on August 19th, its formal acceptance was delayed pending additional tests required by Space Division. The IUS was mated to the launch vehicle on 1 September 1982, and it was mated to the vehicle's DSCS II/III payload on September 29th. Acceptance testing was completed on October 2nd, and the vehicle was prepared for launch.[55]

Launch Vehicle D-01's first Launch CST was aborted on 20 October 1982, but its second Launch CST was completed successfully on October 21st. The countdown was picked up smoothly on October 29th at 2055Z (Greenwich Mean Time), and the first TITAN 34D lifted off Complex 40 at 0405:01Z on 30 October 1982. The TITAN's flight was virtually flawless, and the IUS placed both DSCS satellites into near-perfect equatorial orbits. With the completion of this first highly successful launch operation, the Cape moved solidly into the TITAN 34D era.[56]

Figure 69: First TITAN 34D Launch Vehicle at Complex 40

The next TITAN 34D core vehicle (34D-10) was equipped with a transtage, and it arrived at the Cape on 14 October 1982. It was erected in Cell #2 of the VIB by 12 November 1982, and its Acceptance CST was completed on 14 February 1983. Following hardware acceptance on March 2nd, the vehicle remained in storage in the VIB through 10 July 1983. Core Vehicle D-10 was moved to the SMAB on July 11th. Following the core/solid rocket motor mate, Launch Vehicle D-10 was moved out to Complex 40 on 18 July 1983. The countdown was picked up at 1948Z on January 30th, and the vehicle was launched at 0308:01Z on 31 January 1984. Due to the classified nature of the mission, we may only note that: 1) the countdown was performed without any unscheduled holds, and 2) all flight systems (e.g., electrical, guidance, flight controls, hydraulics, ordnance, solid and liquid propulsion, and payload fairing) performed properly. The launch was successful.[57]

The third TITAN 34D launched from Complex 40 arrived at the Cape on 2 March 1983. That core vehicle (D-11) was equipped with a transtage, and it was erected in Cell #1 of the VIB between 14 and 17 March 1983. Following a successful Acceptance CST on June 24th, the vehicle was accepted formally on 12 July 1983. The vehicle was placed in storage in Cell #2 until it was moved to the SMAB on 10 February 1984. After the core vehicle was mated to its solid rocket motors, Launch Vehicle D-11 was moved out to the launch pad on 18 February 1984. During a Baseline CST at the pad, a Stage I engine squib firing circuit fired late due to a faulty resistor pack, and all resistor packs of the same type had to be replaced with a newer type of resistor pack on the core vehicle and payload fairing. That problem aside, the payload was mated to the vehicle shortly thereafter. Following a successful Launch CST, the vehicle was readied for its classified mission. The countdown on April 14th went well, and there were no unscheduled holds. Launch Vehicle D-11 lifted off the pad at 1652:02Z on 14 April 1984, and all flight systems performed properly. The launch was a success.[58]

One more TITAN 34D/transtage vehicle was launched from the Cape before TITAN launch operations ground to a halt in April 1986. That launch vehicle (D-13) arrived at the Cape on 30 August 1983. The core vehicle was erected in Cell #1 of the VIB by 15 September 1983, but D-13's acceptance testing was delayed pending replacement of its first stage in the early months of 1984. Subsystem testing finally got underway on 3 May 1984, and the Acceptance CST was completed on May 31st. Core Vehicle D-13 was formally accepted on June 8th, and the vehicle was moved to the SMAB for its solid rocket mate on 14 June 1984. Launch Vehicle D-13 was moved to Complex 40 on June 19th, but a schedule change prompted a return trip to the SMAB on July 14th. The vehicle was demated there by the end of July, and it was moved back into storage in Cell #1 of the VIB. The vehicle was reconfigured for a new battery in September 1984, and a Reacceptance CST was accomplished in October. Following the core vehicle/ solid rocket remate at the SMAB, Launch Vehicle D-13 was moved out to Complex 40 and prepared for its classified mission. The countdown was picked up at 1702Z on December 21st, and it proceeded normally to vehicle lift-off at 0002:03Z on 22 December 1984. All flight systems performed properly, and the launch was successful.[59]

Figure 70: Colonel Charles A. Kuhlman

Figure 71: Colonel Dominick R. Martinelli

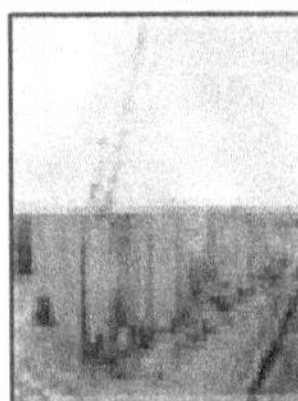

Figure 72: NDT Facility under construction

Figure 73: Raising a Component

Figure 74: Lowering a Component into an X-Ray Cell

All TITAN 34D launch operations at Vandenberg and the Cape were suspended following the TITAN 34D-9 launch failure in April 1986, but it would be wrong to conclude that the suspension allowed the 6555th Aerospace Test Group and the Air Force's TITAN contractors to lapse into a period of inactivity. On the contrary, the space launch recovery effort and TITAN IV program initiatives kept the Test Group's agenda full. Under the command of Colonel Dominick R. Martinelli, the Test Group supervised the initial recovery effort at the Cape. As part of that program, a Non-Destructive Testing (NDT) X-Ray facility was constructed in the ITL Area for the purpose of inspecting TITAN solid rockets for flaws in propellant, restrictors, insulation and podding compounds. Construction of the NDT facility began on 1 October 1986, and solid rocket motor testing was conducted there as part of the TITAN 34D recovery effort from 23 December 1986 through 12 June 1987. During that period, 29 TITAN components were x-rayed on a 24-hour per day schedule. Other non-destructive tests (e.g., ultrasonic testing, laser imaging, thermography and visual inspection with feeler gauges) were accomplished in the ITL Area's Receive and Inspect (RIS) storage building. Rounding out the recovery program, over one hundred TITAN core vehicle components were sent back to Martin Marietta or other vendors for vibration, thermal and acoustic shock testing. The initial NDT effort was concluded in June 1987, but the facility was used later to test other components.[60]

TITAN and Shuttle Military Space Operations

TITAN IV Program Activation and Completion of the TITAN 34D Program

Thanks to the Air Force's initial commitment to the TITAN IV program, Martin Marietta was also in the midst of a $56,900,000 renovation project for the Cape's ITL Area facilities in 1986 and 1987. Under an Air Force contract awarded in February 1985, Martin had hired approximately 50 local area subcontractors to refurbish Launch Complex 41's Mobile Service Tower (MST), its Umbilical Tower (UT) and other facilities in preparation for the new TITAN IV launch vehicle. The VIB, SMAB and other ITL facilities would also be modified and upgraded to support TITAN IV launch operations. Work on Complex 41 began in January 1986 and continued into 1988. The SMAB was modified between November 1986 and the spring of 1987, and work on the VIB was completed between March and October 1987. Work on Transporters #1 and #4 got underway in September 1986 and continued well into 1988.[61]

By the middle of 1988, the first phase of the Cape's TITAN IV program activation was well on its way to completion. In October and November 1987, Complex 41's MST was moved several times to test its operation. By late April 1988, Martin Marietta was ready to turn the MST over to the Air Force for routine maintenance. This is not to say that problems did not remain to be resolved in 1989 and 1990-corrosion was discovered on the recently-refurbished UT in the fall of 1987, and there were a large number of "punchlist" items remaining to be resolved by Martin Marietta and its primary subcontractor, Sauer Construction. Nevertheless, processing of the first TITAN IV hardware began in January 1988, and Martin was already suggesting that Launch Complex 40 be modified to accept TITAN IV vehicles as well as the company's new commercial TITAN III vehicles.[62]

In addition to changes in existing TITAN facilities, two new acquisitions were added to the Test Group's list of TITAN resources as part of the Air Force's commitment to the TITAN IV program: NASA's CENTAUR Payload Operations Control Center (CPOCC) and the ITL Area's Solid Motor Assembly Readiness Facility (SMARF). Following cancellation of the Shuttle/CENTAUR project on 19 June 1986, the CPOCC was transferred to ESMC on 30 January 1987. The Test Group became responsible for the facility, and it began converting the two-story building into a control center for the TITAN IV/ CENTAUR program. The CPOCC designation was deleted, and the facility became simply "Building 27200" until it was renamed the Launch Operations Control Center (LOCC) in 1988. The Test Group had the second floor redesigned to monitor DELTA II operations as well as TITAN IV/CENTAUR operations, and it installed a Launch Management Control Center (LMCC), Command Management Control Center (CMCC) and Spacecraft Control Center (SCC) on that floor. The first floor of the LOCC was designed to serve as a nerve center for TITAN IV/CENTAUR operations in the 1990s.[63]

The Test Group's other new TITAN facility was drawn on a clean sheet of paper. In 1988, Martin Marietta proposed the SMARF as the best of three alternatives for supporting the TITAN IV's new Solid Rocket Motor Upgrade (SRMU) design. The 320 x 185 x 250-foot facility would need a building site, and the old hydrazine tank car storage area, north of the SMAB, was proposed. The Air Force accepted both of those recommendations in the latter part of 1988, and it got together with the U.S. Army Corps of Engineers to finalize the floor plan and stacking cell platform elevations in February 1989. Following the final 100 Percent Design Review in September 1989, more than 50 prospective contractors attended the SMARF site visit and pre-proposal conference at the Cape on 11 October 1989. The contract award was delayed for several months due to the facility's probable threat to the Florida Beach Mouse (a recent addition to the endangered species list), but the $39,976,984 contract for the project was awarded to Frank J. Rooney, Inc. on 18 April 1990. Site preparation was underway by the end of May 1990, and the SMARF was essentially complete by the end of 1991. Unfortunately, crane problems plagued the SMARF in 1992, and they continued to delay operations in 1993.[64]

Figure 75: Launch Operations Control Center (LOCC)

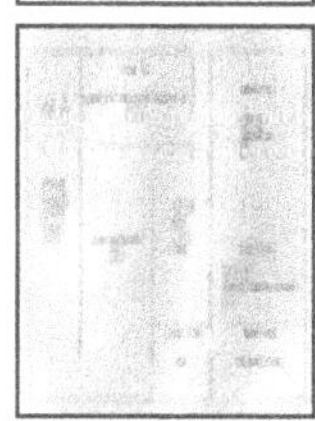

Figure 76: LOCC first floor

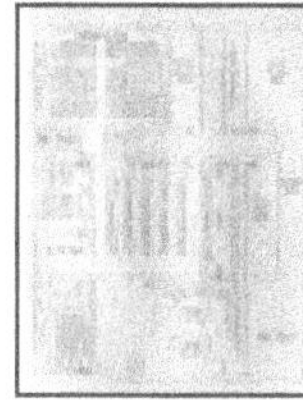

Figure 77: LOCC second floor

Figure 78: SMARF under construction in January 1991

As construction and renovation at the Cape continued, the Test Group prepared for the return of TITAN 34D launch operations and the debut of TITAN IV launch operations. Vandenberg had the honor of launching the first post-recovery TITAN 34D mission on 26 October 1987, but the next TITAN 34D mission was launched from the Cape on 29 November 1987. The launch vehicle (D-8) for that classified mission was equipped with a transtage, and it flew from Complex 40 almost three years after it arrived at the Cape. The countdown was picked up at 1849:00Z on November 28th. Despite two unscheduled holds, the final count went smoothly. The vehicle lifted off Complex 40 at 0327:20Z on November 29th. All flight systems performed properly, and the launch was a complete success.[65]

Figure 79: SMARF completed in December 1991

The next TITAN 34D launch at the Cape involved a classified payload and Launch Vehicle D-3. Like Launch Vehicle D-8, D-3 was equipped with a transtage, and its launch had been held up by the TITAN 34D recovery effort in 1986 and 1987. Stages I and II of Core Vehicle D-3 arrived at the Cape on 7 August 1985, and Stage III arrived on 4 September 1985. The core vehicle was erected on Transporter #3 in Cell #1 on 7 September 1985, and the Acceptance CST was completed on 7 February 1986. The vehicle remained in storage on its transporter in the VIB from March 1986 to April 1987. Stages II and III were disassembled so Stage II could be sent back to Martin Marietta for an engine upgrade on 23 April 1987. Stage II was upgraded, and it was returned to the Cape on 28 July 1987. Following re-erection of the core vehicle on August 5th, power was restored to D-3 on 15 January 1988 for extensive retesting in the VIB. A Retest CST was run on 11 February 1988, and the vehicle was moved to the SMAB for solid rocket mating on February 17th. Launch Vehicle D-3 was moved out to Complex 40 on 20 February 1988. Testing continued at the launch site over the next six months. The countdown for the launch was picked up at 0405Z on 2 September 1988. Lift-off was recorded at 1205:02Z, and the first part of the flight went well. Unfortunately, the transtage failed to boost the payload to mission orbit. Subsequent investigations by Space Systems Division and the TITAN 34D Program Office could not pinpoint the exact cause of the anomaly, but a highly technical analysis of telemetry data revealed that the transtage lost pressure through a small hole in its fuel tank feed system before the transtage's first burn. The investigators concluded that there had been enough pressure for the transtage's first burn, but not enough pressure for its second burn. Space Systems Division's two remaining TITAN 34D/Transtage vehicles were thoroughly inspected to prevent a similar occurrence in the future.[66]

As TITAN 34D launch operations continued, Colonel Robert B. Bourne assumed command of the 6555th Aerospace Test Group near the beginning of the TITAN IV era at the Cape. The first TITAN IV liquid rocket engines were installed on the TITAN IV "pathfinder" vehicle at the end of January 1988, shortly before the core vehicle was erected in the VIB. Four TITAN IV solid rocket motor segments were received at the SMAB by the middle of February 1988, and two electrical functional tests were conducted in early March. As "bugs" were worked out of various systems, the core vehicle had its first successful CST on 11 May 1988. The vehicle was moved to the SMAB around the middle of May. Following a successful mate with two five-segment stacks of solid rocket motor segments, the pathfinder vehicle was moved out to Complex 41 on Saturday, May 21st. The sixth segment was added to both stacks on May 22nd, and the seventh segment was added to the "north" stack on May 23rd. Communications problems and high winds over the Cape delayed IUS mating for a short time, but the IUS was mated to the vehicle on 16 June 1988. A dummy TITAN IV spacecraft was mated to the IUS a few days later, but the Baseline CST was attempted three times before it was completed successfully on July 12th. The payload fairing was mated to the vehicle during the first week in August, and electromagnetic compatibility tests were completed on 23 August 1988.[67]

The Launch Readiness Verification (LRV) for the pathfinder was attempted twice in September 1988, but lousy weather, fire alarm indications and a balky Mobile Service Tower stymied those efforts. The most serious obstacle was the MST: the tower could not be jacked up high enough to clear the pad's launch piers, and further investigation revealed cracks in three strongbacks supporting the jacking system. The LRV was suspended while the strongbacks were repaired, tested and repainted. The MST was exercised successfully on 25 November 1988. A bridge crane required adjustments in December 1988, and Complex 41's Umbilical Tower and 1750-kilowatt generator were out of service at the end of March 1989. Those problems aside, preparations for the first TITAN IV launch continued.[68]

Figure 80: TITAN 34D Core Vehicle Erection

Figure 81: TITAN 34D Solid Rocket Motor Mate

Figure 82: First TITAN IV Core Vehicle arrives at the Skid Strip January 1988

Figure 83: Capping off a Solid Rocket Motor at the Pad

Figure 84: Colonel Robert B. Bourne

The first TITAN IV vehicle supported a classified mission. Its launch had been scheduled for 7 June 1989, but the lift-off was pushed to June 14th due to a range timing generator problem and a computer malfunction. The countdown was picked up at 0254Z on June 14th. Two unscheduled holds were called to let the launch team catch up on checklist items that were behind schedule, and another hold was called for a high temperature reading on the vehicle's S-Band transmitter. Following the last unscheduled hold, the countdown proceeded uneventfully, and the TITAN IV lifted off Complex 41 at 1318:01Z on 14 June 1989. While the launch was successful, an anomaly occurred approximately 130 seconds into the flight: a failure in a thrust chamber regenerative tube caused one of the first stage's two liquid rocket nozzles to gimbal to the limit of its travel. Luckily, there was sufficient control in the other engine to maintain the launch vehicle's stability, and the flight was successful.[69]

Though the first TITAN IV mission ended well, Space Systems Division was very concerned about the TITAN IV-1 flight anomaly. The TITAN IV Program Office, Martin Marietta, the rocket engine manufacturer (Aerojet Tech Systems) and the Aerospace Corporation conducted an exhaustive investigation of the incident. A component defect was not ruled out, but the investigators concluded that the most probable cause for the failure was contamination in the tubes composing the combustion chamber nozzle. The contamination probably restricted the cooling flow of fuel to the tubes, and one of the tubes burned out as a result. In the short term, contractors intended to pay more attention to signs of contamination. In the long term, design modifications, better process control and more non-destructive testing were planned to detect and eliminate all potential sources of contamination.[70]

As the Air Force closed out its TITAN 34D program in favor of the TITAN IVs, the last two TITAN 34D flights (D-16 and D-2) were launched from Complex 40 in May and September 1989. Core Vehicle D-16 arrived at the Cape on 28 April 1988, and it was erected on Transporter #3 in May. Unfortunately, the vehicle's transtage had to be sent back to Martin Marietta's plant in Denver, Colorado for some modifications on 12 November 1988. The transtage was returned to the Cape on 24 January 1989. The Acceptance CST was completed on March 10th, and the core vehicle was mated to its solid rocket motors around 20 March 1989. Launch Vehicle D-16 was moved out the Complex 40 shortly thereafter. Following payload mate, fairing installation and a good Launch CST, a launch was attempted on 9 May 1989. It had to be scrubbed due to an "incompatibility" between the Range Sequencer and the TITAN program's new Programmable Aerospace Control Equipment, but the problem was corrected quickly. The countdown was picked up again on May 10th, and it proceeded smoothly until an 11 minute hold was called for weather at T minus five minutes. The countdown resumed, and TITAN 34D-16 lifted off with its classified payload at 1947:01Z on 10 May 1989. According to the Martin Marietta test report on the vehicle, all TITAN systems performed normally in flight.[71]

Figure 85: First TITAN IV launch from Complex 41
14 June 1989

The last TITAN 34D launched from the Cape had an extensive processing history between the time it first arrived at the Cape (e.g., 19 August 1981) and the time it was erected for the final time on Transporter #3 in Cell #1 on 13 December 1988. In the last series of processing operations, the core vehicle (D-2) was transferred to storage in Cell #4 on 16 December 1988. Core Vehicle D-2 was moved back to Cell #1 after the transtage arrived in mid-March 1989. The transtage was erected on the core vehicle on 28 March 1989. The Acceptance CST was completed successfully on 23 June 1989. Core Vehicle D-2 was moved to the SMAB for solid rocket mating on July 2nd, and Launch Vehicle D-2 was moved out to Complex 40 on 5 July 1989. The vehicle was mated to a classified payload and prepared for launch. Though the first Launch CST failed on August 21st, the Launch CST on August 27th was completely successful. A balky Mobile Service Tower delayed pre-launch activities on September 4th, but a 22-minute-long user hold brought operations up to speed at T minus 30 minutes. After the countdown resumed at 0524Z, it proceeded without incident to vehicle lift-off at 0554:01Z on 4 September 1989. Once again, Martin Marietta's test report confirmed that all TITAN systems performed normally in flight.[72]

Figure 86: Last TITAN 34D launch from Cape Canaveral
4 September 1989

TITAN and Shuttle Military Space Operations

TITAN IV Operations after First Launch

With regard to later TITAN IV launch operations at the Cape, the second TITAN IV was a basic launch vehicle with no upper stage. It was referred to as the TITAN IV/NUS or Vehicle K-4. The core vehicle arrived at the Cape on 29 June 1989, and it was erected on Transporter #1 in Cell #2 by 25 October 1989. The Acceptance CST was completed on 25 January 1990, and the core vehicle was mated to its partial solid rocket motor stacks on March 1st. Launch Vehicle K-4 was moved out to Complex 41 on 12 March 1990, and solid rocket motor stacking was completed at the pad on April 28th. A classified payload was mated to the launch vehicle on May 5th, and the Launch CST was completed successfully on May 30th. The countdown for the Cape's second TITAN IV launch was largely uneventful (i.e., only one six-minute unscheduled hold for weather constraints was noted). The vehicle lifted off the pad at 0521:41Z on 8 June 1990. Martin Marietta's test report on the vehicle confirmed that all TITAN systems performed normally in flight.[73]

The third TITAN IV vehicle (K-6) arrived at the Cape on 10 December 1989. The core vehicle was erected on Transporter #1 in Cell #2 by 24 February 1990, and the Acceptance CST was completed successfully by June 19th. The core vehicle was moved to the SMAB on 22 June 1990, and it was mated to its solid rocket motors on June 24th. Launch Vehicle K-6 was moved out to Complex 41 on 25 June 1990, and the IUS was mated to the launch vehicle on July 6th. The classified payload was mated on 25 July 1990, and the payload fairing was installed by September 6th. The Launch CST on September 9th was successful, but the first launch attempt was scrubbed on September 23rd due to range computer software loading difficulties. Another Launch CST ran successfully on November 5th, and the countdown on November 12th was largely uneventful. TITAN IV-6 lifted off Complex 41 at 0037:02Z on 13 November 1990. According to Martin Marietta's test report on the flight, all TITAN systems performed normally.[74]

Figure 87: Arrival of CENTAUR Upper Stage at Skid Strip

Martin Marietta began processing TITAN IV/CENTAURs in the early 1990s, but TITAN IV/ CENTAUR launch operations were delayed by two ATLAS I/ CENTAUR upper stage flight failures in 1991 and 1992. In effect, the flight mishaps cast doubt on the reliability of CENTAUR upper stages in general, and all ATLAS/CENTAUR and TITAN/CENTAUR launch operations had to stand down (albeit for varying periods of time). In the aftermath of the failures, the Air Force agreed to let Martin Marietta roll back and destack two different TITAN IV vehicles that had been standing (or were likely to stand) on Complex 41 for more than a year. In the first instance, the TITAN IV languished on the pad during the lengthy investigation into the first ATLAS I/CENTAUR flight failure. That TITAN had been standing on the pad for about a year when Martin recommended it be rolled back and demated in July 1992. In the aftermath of the second investigation, TITAN/CENTAUR upper stages had to be modified. Since CENTAUR modifications pushed the first TITAN/CENTAUR launch back to the summer of 1993 (at the very least), the next TITAN IV/CENTAUR (stacked in the summer of 1992) had to be rolled back from the pad and destacked in February 1993. Another TITAN IV/CENTAUR was moved out to Pad 41 on 30 March 1993, but it remained to be seen if it would suffer the same fate as its immediate predecessors.[75]

Despite the setback in TITAN IV/CENTAUR operations, the Air Force continued to upgrade other TITAN facilities at the Cape to handle the TITAN IV vehicle. In the spring of 1990, Martin Marietta was awarded a major contract to upgrade Complex 40 into a TITAN IV launch site. Martin Marietta, in turn, awarded a $100 million subcontract to Bechtel National, Inc. in early June 1990 to cover the remaining design, procurement and construction effort. The project included a new Mobile Service Tower (MST), a new Umbilical Tower (UT) and supporting systems. Both new towers were designed with a high level of corrosion resistance in mind, and the UT was built stronger to handle the heavier SRMU loads anticipated in the near future. The old UT was completely demolished by the end of August 1990, and MST demolition was ahead of schedule. Reconstruction was a monumental task: the new 265-foot-tall MST contained 21 working levels and the 170-foot-tall UT contained 15 working levels. A new Air-Conditioning Shelter (ACS) was equipped to handle all projected requirements for the new TITAN IV, its CENTAUR upper stage and payloads. Launch pad modifications extended to fuel and oxidizer waste tank containment areas, security systems and the fuel vapor incinerator area.[76]

Figure 88: Erection and Mate of CENTAUR at VIB Cell 3 December 1990

The new UT was completed in the fall of 1991, and the new MST was completed in April 1992. All aerospace ground equipment was installed by the spring of 1992, but the launch facility experienced crane test failures in April, and electromagnetic interference (EMI) leaks were discovered around Complex 40's EMI doors. The leaks were repaired, but four doors failed independent EMI tests in August 1992. Martin Marietta was compelled to develop recovery plans for the EMI condition, which became the "driver" in delaying completion of the project. By mid-October, the problem pushed Complex 40's initial TITAN IV launch capability to 1 December 1992. New "finger stock" passed high frequency and low frequency tests in November 1992, and that stock was installed on 55 EMI doors in December 1992. Air-Conditioning Shelter enhancement continued into the early part of 1993, and Complex 40 received its first TITAN IV/CENTAUR launch vehicle on 2 June 1993. That vehicle was launched successfully on a MISTAR communications satellite mission on 7 February 1994.[77]

Figure 89: Reconstruction of Complex 40
May 1991

TITAN and Shuttle Military Space Operations

Space Shuttle Military Missions

This chapter would not be complete without a summary of all the major military missions carried out aboard the Space Shuttle in the 1980s and early 90s. A short review of Shuttle payload processing procedures and range support responsibilities is pertinent to this discussion, but the Test Group's role in setting up military Shuttle operations in the 1970s and early 80s really needs to be mentioned first. As we noted earlier, the 6555th Aerospace Test Group established its Space Transportation System (STS) Division on 1 July 1974. The Division was created to ensure that Defense Department requirements were included in plans for future Shuttle operations at the Kennedy Space Center (KSC). As two of its earliest accomplishments, the Division got NASA to agree to the Defense Department's requirement for vertical payload installations at the Shuttle launch pad and a secure conference area in the Firing Room of the Shuttle Launch Control Center (LCC). Under the direction of Lt. Colonel George L. Rosenhauer, the Division continued to serve as an intermediary between KSC and the Defense Department payload community. The Division not only gave the payload community a better understanding of schedule and contractual constraints affecting KSC ground operations, it also gathered a more detailed set of requirements from military payload programs to help NASA support those programs. The Division also helped the 6595th Space Test Group develop requirements for a Shuttle Launch Processing System at Vandenberg Air Force Base.[78]

On 8 June 1977, Lt. Colonel Warren G. Green (Chief, Space Launch Vehicle Systems Division) succeeded Lt. Colonel Rosenhauer as Chief of the STS Division. By that time, the Division was actively engaged in planning IUS ground support operations including IUS processing operations at the SMAB. In October 1977, three IUS officer positions were transferred from the Space Launch Vehicle Systems Division to the STS Division to help support the growing IUS workload. In December, the Division participated in the 90 Percent Design Review for the Payload Ground Handling Mechanism (PGHM). The PGHM would be used to support and install all payloads destined for vertical integration at the Shuttle's two launch pads. The Division also provided selection criteria and background information to help the Space and Missile Systems Organization select its Shuttle payload integration contractor. Martin Marietta was awarded the Shuttle payload integration contract on 15 September 1977.[79]

As preparations for military Shuttle operations continued, the STS Division identified and analyzed many problems associated with "factory-to-pad" processing of military payloads. The Division's findings helped justify the need for an off-line Shuttle Payload Integration Facility (SPIF), and they convinced the AFSC Commander to approve the SMAB's west bay as the site for the SPIF in January 1979. Pooling their expertise, the STS Division and the Space Launch Vehicle Systems Division conducted a comprehensive review of the SPIF's requirements in July 1979. As work on the SPIF got underway, the 6555th Aerospace Test Group formed the STS/IUS Site Activation Team in September 1981 to address problems associated with the first IUS processed aboard the Shuttle. The new team reinforced rather than diminished the STS Division's role in STS operations, and a new branch was added to the Division in August 1982 to take advantage of "lessons learned" from the first military payload processed aboard the Shuttle. The new branch provided officers to NASA's Space Vehicle Operations Directorate to help streamline military payload processing at the launch base. As we noted earlier, the STS Division and the Satellite Systems Division were consolidated to form the Spacecraft Division on 1 November 1983.[80]

The first military Shuttle mission was launched from Pad 39A at 1500Z on 27 June 1982. Military space missions also accounted for part or all of 14 out of 37 Shuttle flights launched from the Cape between August 1984 and July 1992. While many details of those missions are not releasable, some features of Shuttle payload ground processing operations and range support requirements can be summarized for what might be termed a "typical" military space mission. One process common to many military Shuttle missions was the preparation of the Inertial Upper Stage (IUS). Though the ultimate destination of the IUS was mission-specific, the IUS was processed in one of two basic assembly/checkout flows (i.e., one for military payloads and the other for NASA spacecraft). Before either process began, the Inertial Upper Stage's structural assemblies, avionics and flight batteries were received at hangars E and H and placed in various storage areas at the Cape. At the appropriate time, all vehicle elements were transferred to the SMAB, where they were assembled and checked out. Following power up checks and functional testing, the military IUS was cleaned and transferred to the SPIF. For civilian missions, IUSs entered a different assembly/checkout flow at this point in the process. They were sent directly to NASA's Vertical Processing Facility on Merritt Island.[81]

Following its arrival at the SPIF, the military IUS was placed in an integration cell. The IUS was mated to a military spacecraft at that time, and the IUS and spacecraft interfaces were checked to ensure everything matched up correctly. During the next major step in the process, the Orbiter Functional Simulator (OFS) was used to verify the upcoming mate with the Shuttle orbiter, and the Vertical Integration Building's Checkout Station (VIB/COS) controlled the IUS during that procedure. Once that step was completed, the spacecraft and IUS were placed in a NASA canister and moved from the SPIF to the Rotating Service Structure (RSS) at the launch pad on Merritt Island. The payload was attached to the Payload Ground Handling Mechanism, the active Safe and Arm mechanism was installed (but not connected electrically), and ordnance circuits were completed on the destruct mechanism. After those tasks were completed, the payload was placed in the orbiter for IUS and payload/orbiter launch readiness testing. Following the launch readiness tests, the Safe and Arm mechanism got its electrical connections, and the safing pins on the destruct mechanism were removed. Final IUS preparations were accomplished just before the Shuttle's terminal countdown sequence. The VIB/COS continued to monitor the IUS' functions at the launch pad through Shuttle lift-off.[82]

The Defense Department's range support for Shuttle flights was extensive, and it applied to civilian as well as military missions. The 45th Weather Squadron provided around-the-clock weather forecasts as the launch drew near. Missile Flight Control (45 SPW/SEO) provided the officers responsible for the moment-to-moment safety of the Shuttle's flight from lift-off to orbit. The Eastern Range acted as "lead range" for Shuttle missions, and it provided the lion's share of instrumentation coverage for the Shuttle during the critical boost phase of the mission. Worldwide instrumentation coverage was also provided by the Kwajalein Missile Range, the Western Range, the Pacific Missile Test Center, the Air Force Flight Test Center and the White Sands Missile Range. The DOD Manager for Space Transportation System Contingency Operations maintained a support office (DDMS) at Patrick Air Force Base to serve as a single point of contact for all Shuttle contingency support operations. During Shuttle missions, DDMS staffed a Support Operations Center (SOC) at the Cape to maintain contact with contingency support forces worldwide. Military rescue forces were stationed at Transoceanic Abort Landing (TAL) sites in Africa and Spain. Shuttle contingency forces at Patrick placed three military HH-3E helicopters (complete with aircrews, medical personnel and pararescue specialists) on alert at the Shuttle Landing Facility (SLF) at KSC for every Shuttle mission. Forces from the Air Force Reserve, the National Guard, U.S. European Command, U.S. Air Forces Europe, the Coast Guard and the Navy were positioned to support an astronaut bailout during the launch phase of each Shuttle mission. More rescue forces stood ready at White Sands Space Harbor, New Mexico to help with launch or landing emergencies. Put simply, Shuttle operations would have been impossible without the Defense Department's multi-faceted support.[83]

Figure 90: Shuttle Pads on Merritt Island 1985

All Shuttle missions had basic flight events in common. At lift-off, the orbiter's three main engines fired in concert with two solid rocket boosters to produce approximately 6,500,000 pounds of thrust at lift-off. The vehicle rolled over into its flight trajectory shortly after lift-off, and it retained its launch configuration until the solid rocket boosters separated approximately two minutes into the flight. At about T plus 520 seconds, the orbiter's main engines shut down, and the Shuttle separated from its external tank. The Shuttle crew prepared for orbital insertion, and the vehicle's Orbital Maneuvering System (OMS) fired two engines to accelerate the Shuttle into an initial, elliptical orbit. Once the OMS engines circularized the Shuttle's orbit, the orbiter's Reaction Control System (RCS) maintained it. The RCS thrusters also allowed the Shuttle to maneuver in all directions while it orbited Earth.[84]

Once on-orbit operations were completed, the crew prepared the Shuttle for reentry and landing. The RCS was used to turn the orbiter around and upside down. In that attitude, the OMS engines were fired to slow the vehicle and permit its safe reentry into Earth's atmosphere. Once the Shuttle's velocity was diminished sufficiently, the RCS was fired again to turn the vehicle around and put it in the typical "nose-up" position for reentry. As the Shuttle reentered the atmosphere, a communications blackout was produced by surface friction and heat ionization. The blackout lasted for about two minutes. Landing occurred about half an hour later at one of the runways at Edwards Air Force Base or the Shuttle Landing Facility at KSC.[85]

With this background in mind, we are ready for a brief review of the Shuttle program's military space operations. While details concerning the nature of the first Shuttle/DOD payload remain classified, we may note that it arrived at the Cape in April 1982. It was processed by an Air Force/NASA/contractor team, and it was loaded aboard the Shuttle Columbia as the vehicle stood on Pad 39A. Following an 87-hour countdown, Columbia lifted off at 1500:00Z on 27 June 1982. Navy Captain Thomas K. Mattingly, II and Air Force Colonel Henry W. Hartsfield, Jr. conducted the military mission in addition to several civilian experiments while "on-orbit," and the long-term effects of temperature changes on Shuttle subsystems were studied along with a survey of orbiter-induced contamination in the Shuttle's payload bay. Columbia made a hard runway landing at Edwards Air Force Base at 1609:00Z on 4 July 1982.[86]

The first of five operational SYNCOM IV military communications satellites was launched on Discovery's maiden flight on 30 August 1984. The flight supported a mixed DOD/civilian mission, and Discovery's on-orbit agenda included the deployment of two civilian satellites (e.g., AT&T's TELSTAR 3-C and Satellite Business Systems' SBS-D) and a solar array experiment (OAST-1). Colonel Henry W. Hartsfield, Jr. commanded Discovery on the mission, and Navy Commander Michael L. Coats piloted the orbiter. Lt. Colonel Richard M. Mullane, Dr. Judith A. Resnick and Dr. Steven A. Hawley served as mission specialists. Mr. Charles D. Walker was the payload specialist. The launch from Pad 39A had been scrubbed on 25 June 1984 due to a Shuttle computer malfunction, and computer software problems pushed the rescheduled lift-off from August 29th to August 30th. The countdown on August 30th included one unscheduled hold for an unidentified aircraft that intruded into the launch area, but Discovery's lift-off at 1241:50Z was untroubled. All three satellites were deployed successfully during the flight, and Discovery landed on Runway 17 at Edwards Air Force Base at 1338:00Z on 5 September 1984.[87]

Mission 51-A was Discovery's second voyage into space, and it featured a military spacecraft among its payloads. The primary objectives of the mission were to: 1) deploy the second operational SYNCOM IV satellite and TELESAT CANADA's ANIK D2 commercial communications satellite, 2) retrieve two commercial satellites (e.g., PALAPA B2 and WESTAR VI) from useless orbits and 3) conduct a variety of biological experiments. Navy Captain Frederick H. Hauck commanded Discovery on the mission, and Navy Commander David M. Walker piloted the orbiter. The mission specialists were Dr. Joseph P. Allen, Dr. Anna L. Fisher and Navy Commander Dale A. Gardner. The lift-off was scheduled for 7 November 1984, but upper level wind shear delayed the launch until November 8th. Discovery was launched from Pad 39A at 1215:00Z on 8 November 1984. The ANIK D2 satellite was deployed successfully at 2104Z on November 9th, and the military payload-SYNCOM IV-was deployed successfully at 1256Z on November 10th. The rendezvous and satellite capture sequences were completed successfully over the next four days in space, and Discovery landed at KSC's Shuttle Landing

Facility at 1200:01Z on 16 November 1984.[88]

Though the first all-military Shuttle mission was originally scheduled for launch on 8 December 1984, it did not lift off until 24 January 1985. Captain Thomas K. Mattingly, II was selected to command Discovery on the highly classified mission. The orbiter was piloted by Air Force Colonel Loren J. Shriver, and the mission specialists were Air Force Major Ellison S. Onizuka and Marine Corps Lt. Colonel James F. Buchli. Air Force Major Gary E. Payton served as payload specialist. The launch was delayed on January 23rd due to weather, and cold weather held up cryogenic fueling operations for two hours on the 24th. Those delays aside, the last four hours of the countdown proceeded smoothly, and Discovery lifted off Pad 39A at 1950:00Z on 24 January 1985. Details of the mission are not releasable. Discovery landed at KSC at 2123:24Z on 27 January 1985.[89]

Figure 91: SYNCOM IV Satellite

Figure 92: DISCOVERY lift-off, Pad 39A
30 August 1984

Figure 93: DISCOVERY lifts off Pad 39A on the first all-military Shuttle mission
24 January 1985

The third SYNCOM IV spacecraft was deployed along with TELESAT CANADA's ANIK-C satellite during Discovery's mission in mid-April 1985. Air Force Colonel Karol J. Bobko served as Shuttle commander for the mission, and Navy Captain Donald E. Williams piloted the orbiter. The mission specialists were Dr. M. Rhea Seddon, Mr. S. David Griggs and Dr. Jeffrey A. Hoffman. The payload specialists were Mr. Charles D. Walker and Senator E. Jake Garn. There were two unscheduled holds during the countdown on April 12th, but the terminal count was uneventful, and Discovery lifted off Pad 39A at 1359:05Z on 12 April 1985. Discovery's crew deployed the ANIK-C satellite successfully on the first day of the mission, and the SYNCOM IV was deployed on Day 2. Unfortunately, the SYNCOM IV's perigee kick motor failed to fire, and two more days were added to the mission to allow a rendezvous and an improvised restart of the spacecraft. Two "flyswatter" devices were attached to the Shuttle's Remote Manipulating System (RMS) to allow the crew to depress the SYNCOM IV's timer switch. Despite a successful rendezvous and a switch reset on Day 6, the attempt failed. The SYNCOM

IV spacecraft was left in orbit to be retrieved and redeployed in early September 1985. Discovery landed at KSC's Shuttle Landing Facility at 1355:37Z on 19 April 1985.[90]

Discovery's sixth trip into space was launched in late August 1985. It was designed to: 1) deploy three communications satellites and 2) retrieve, repair and redeploy the SYNCOM IV communications satellite that had been stranded in a useless orbit since mid-April 1985. Air Force Colonel Joe H. Engle commanded Discovery on the mission, and Air Force Colonel Richard O. Covey piloted the orbiter. The mission specialists were Dr. James D. Van Hoften, Mr. John M. Lounge and Dr. William F. Fisher. One of the three satellites launched onboard Discovery in August was the fourth operational SYNCOM IV spacecraft. The other two spacecraft were Australia's AUSSAT-1, and the American Satellite Company's ASC-1. The mission was scrubbed for foul weather on August 24th, and another launch scrub was caused by a faulty backup flight system computer on August 25th. The countdown was started again at 0205Z on August 27th, and it proceeded smoothly except for a three-minute extension in a built-in hold to clear traffic in a solid rocket booster retrieval area. Discovery lifted off Pad 39A at 1058:01Z on 27 August 1985. The AUSSAT-1 spacecraft was ejected from the orbiter's cargo bay at 1733Z on the 27th, and the satellite's deployment and perigee kick motor burns were both successful. The ASC-1 deployment and boost were also successful on Day 1 of the mission. The SYNCOM IV-4 deployment went extremely well on Day 3, and Discovery's crew prepared for their rendezvous with the wayward SYNCOM IV-3 spacecraft on Day 5. The spacecraft was retrieved by Van Hoften and Fisher, and they completed their repairs on the satellite on Day 6 of the mission. SYNCOM IV-3 was redeployed at 1512Z on 1 September 1985. Unlike its earlier performance in April, the spacecraft began sending good telemetry data to ground stations shortly thereafter. Discovery landed on Edwards' Runway 23 at 1315Z on 3 September 1985.[91]

The Shuttle Atlantis' maiden flight was completed in early October 1985, and it was dedicated to a highly classified military mission. Colonel Karol Bobko commanded Atlantis on the flight, and Air Force Lt. Colonel Ronald J. Grabe piloted the orbiter. The mission specialists were Army Colonel Robert L. Stewart and Marine Corps Major David C. Hilmers. Air Force Major William A. Pailes served as payload specialist. Details of the mission remain classified, but we may confirm that Atlantis was launched from Pad 39A at 1515:30Z on 3 October 1985. Atlantis landed on Edwards' Runway 23 at 1700Z on October 7th.[92]

The Shuttle's next military mission was put on hold after the Challenger disaster, but it was carried out by Atlantis between 2 and 7 December 1988. The mission was highly classified, so most details are not releasable. The mission was commanded by Navy Commander Robert L. Gibson, and the orbiter was piloted by Air Force Colonel Guy S. Gardner. The mission specialists were Air Force Colonel Richard M. Mullane, Air Force Lt. Colonel Jerry L. Ross and Navy Commander William M. Shepherd. Though the countdown was picked up at 0230Z on December 1st, upper level wind shear effects delayed the launch until December 2nd. The countdown was picked up again on December 2nd, but a problem with a ground feed liquid oxygen valve required a 50-minute unscheduled hold at T minus 180 minutes. Wind shear problems forced another delay at T minus nine minutes for an additional 99 minutes, but the final unscheduled hold (at T minus 31 seconds) only lasted 71 seconds. Atlantis lifted off Pad 39B at 1430:34Z on December 2nd. The Shuttle landed at Edwards Air Force Base at 2336:11Z on 6 December 1988.[93]

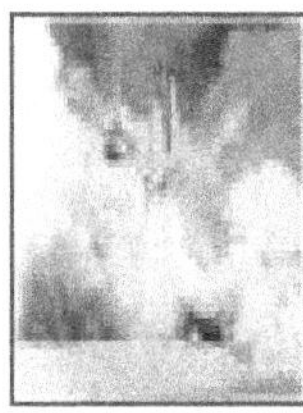

Figure 94: ATLANTIS lifts off Pad 39A on her maiden flight 3 October 1985

After a military furlough of seven years, Columbia was pressed into service to support her second military space mission in August 1989. Once again, the mission was highly classified, so only a few details are releasable. Air Force Colonel Brewster H. Shaw, Jr. commanded Columbia on this all-military mission. Navy Commander Richard N. Richards served as the pilot, and the mission specialists were Navy Commander David C. Leetsma, Army Lt. Colonel James C. Adamson and Air Force Major Mark N. Brown. The countdown got underway on 8 August 1989. A user data link problem delayed the countdown for approximately 70 minutes during a built-in hold, but the count proceeded normally after that incident. Columbia lifted off Pad 39B at 1237:00 on 8 August 1989. In addition to deploying their military payload successfully, Columbia's crew conducted several on-orbit experiments during the five-day mission. The Shuttle landed on Edwards' Runway 22 at 1337Z on 13 August 1989.[94]

Air Force Colonel Frederick D. Gregory commanded Discovery on her second all-military Shuttle mission in late November 1989. Air Force Colonel John E. Blaha was the pilot, and the mission specialists were Dr. F. Story Musgrave, Dr. Kathryn C. Thornton and Navy Captain Sonny Carter. The countdown on November 23rd proceeded uneventfully until T minus five minutes, when a three-minute and thirty-second hold was called to let the user complete checklist items. The countdown resumed, and Discovery lifted off Pad 39B at 0023:30Z on 23 November 1989. Though Discovery's landing was delayed until November 27th due to high winds over Edwards Air Force Base, the Shuttle made a successful landing on Runway 4 at 0030Z on 28 November 1989. Defense forces were released to their local commanders about 30 minutes later, and Discovery was ferried back to KSC via Eglin Air Force Base, Florida on 3 and 4 December 1989.[95]

Columbia's ninth space mission was a mixed military/civilian operation. It was commanded by Navy Captain Daniel C. Brandenstein and piloted by Navy Commander James D. Weatherbee. The mission specialists were Ms. Marsha S. Ivins, Dr. Bonnie J. Dunbar and Dr. G. David Low. The mission had three main objectives: 1) deploy the fifth SYNCOM IV military satellite, 2) retrieve the Long Duration Exposure Facility (LDEF) deployed by the Shuttle Challenger in early April 1984, and 3) conduct a variety of experiments in the Shuttle's middeck area. A launch attempt on 8 January 1990 was scrubbed due to weather, but the countdown on January 9th proceeded smoothly, and Columbia was launched from Complex 39A at 1235:00Z on 9 January 1990. The SYNCOM IV-5 spacecraft was deployed successfully at 1318Z on January 10th, and Columbia rendezvoused with the LDEF on January 12th. All middeck experiments were underway by the end of Day 2 of the mission. Though the Shuttle's landing was delayed a day for weather, Columbia landed safely on Edwards' Runway 22 at 0935:38Z on 20 January 1990. Defense support forces were released to normal operational control about an hour later, and Columbia was ferried back to KSC (with an overnight stay at Kelly Air Force Base, Texas) on 25 and 26 January 1990.[96]

Under the command of Navy Captain John O. Creighton, Atlantis lifted off Pad 39A on another all-military Shuttle mission at 0750:22Z on 28 February 1990. The pilot for the flight was Air Force Colonel John H. Casper, and the mission specialists were Marine Corps Lt. Colonel David C. Hilmers, Colonel Richard M. Mullane, and Navy Commander Pierre J. Thout. Though details of the mission remain classified, the flight was successful. Atlantis landed on Edwards' Runway 23 at 1808:44Z on 4 March 1990. The Shuttle was ferried back to KSC on 10 and 11 March 1990.[97]

Colonel Richard O. Covey commanded Atlantis on another all-military Shuttle mission in November 1990. The pilot for that mission was Navy Commander Frank L. Culbertson, Jr., and the mission specialists were Marine Corps Colonel Robert C. Springer, Air Force Lt. Colonel Carl J. Meade and Army Captain Charles D. Gemar. The launch was originally planned for the summer of 1990, but it was delayed after hydrogen leaks were found in the Atlantis and Columbia orbiters. (Atlantis was rolled back to the VAB for repair toward the end of July 1990.) A new mission execution order (90-7) was implemented on 21 October 1990, and it announced a tentative launch date of 10 November 1990. The countdown was picked up on November 15th at 1340Z, and it proceeded smoothly to a built-in hold at T minus 9 minutes. That hold was extended two minutes and 34 seconds to allow the user to catch up on checklist items, and the countdown proceeded to lift-off at 2348:15Z on 15 November 1990. The mission was highly classified, so on-orbit details are not releasable. Atlantis' crew planned to land at Edwards Air Force Base on November 19th, but strong winds delayed the landing and forced NASA to divert the orbiter to KSC's Shuttle Landing Facility instead. Atlantis landed on KSC Runway 33 at 2142:43Z on 20 November 1990.[98]

Discovery supported a Strategic Defense Initiative (SDI) mission in the spring of 1991. Navy Captain Michael L. Coats commanded the mission, and Lt. Colonel L. Blaine Hammond, Jr. piloted the orbiter. The mission specialists were Mr. Gregory J. Harbaugh, Air Force Colonel Guion S. Bluford, Jr., Air Force Lt. Colonel Donald R. McMonagle, Charles L. Veach and Mr. Richard J. Hieb. The first mission execution order (91-1) was implemented on 13 February 1991, but the intended launch date of 9 March 1991 was abandoned after cracks were found on the hinges of the external tank's umbilical doors in late February 1991. A new mission execution order went into effect on April 1st, and the launch was carried out under that order. The countdown on 28 April 1991 was delayed at T minus 9 minutes for approximately half an hour to correct a flight recorder problem, but Atlantis lifted off Pad 39A safely at 1133:14Z on the 28th.[99]

Discovery's SDI mission featured two deployable payloads, three orbiter bay payloads and two middeck experiments. The Infrared Background Signature Survey (IBSS) was onboard to help define SDI systems and gather infrared data on Shuttle exhaust plumes, Earthlimb and aurora phenomena, chemical/ gas releases and celestial infrared sources. It consisted of two deployable hardware elements (e.g., the Shuttle Pallet Satellite II and a collection of three Chemical Release Observation sub-satellites) and a non-deployable Critical Ionization Velocity element. The Air Force Program 675 payload was included on the mission to gather infrared, ultraviolet and x-ray data on auroral, Earthlimb and celestial sources. It consisted of five experiments mounted on a pallet in the Shuttle payload bay. The Space Test Payload-1 (STP-1) was a secondary payload consisting of five experiments designed to gather data on: 1) fluid management in weightless conditions, 2) MILVAX computer and erasable optical disk performance in weightless conditions, 3) atomic oxygen glow effects, 4) free particles present in the Shuttle payload bay during flight ascent and 5) the upper atmosphere's composition. The Cloud Logic to Optimize Use of Defense Systems (CLOUDS) experiment used a 36-exposure camera to photograph clouds and correlate cloud characteristics with their impact on the efficiency of military surveillance systems. The hand-held Radiation Monitoring Equipment III (RME III) sensor was included on the mission in one of a continuing series of experiments to collect data on gamma radiation aboard the Shuttle.[100]

With Discovery safely in low-Earth orbit, the crew set about completing the mission. The SPAS II was deployed at 0928Z on 1 May 1991. Though problems with the onboard sun sensor forced cancellation of the first exhaust plume observation, other observations went well later in the day. NASA was reportedly "very pleased" with the results. The AFP-675 payload's experiments went well, and 31 of 33 individual experiments were completed by the time the Shuttle's Remote Manipulating System retrieved the SPAS II at 1445Z on May 3rd. Following another day of Earth observations, the SPAS II was returned to the payload bay and stowed. Discovery's deorbit burn occurred around 1750Z on May 6th, and the Shuttle landed at KSC's Runway 15 at 1855Z on the same day.[101]

The last military Shuttle mission before July 1992 was flown by Atlantis. It was commanded by Colonel Frederick D. Gregory and piloted by Air Force Colonel Terence T. Henricks. The mission specialists were Dr. F. Story Musgrave, Navy Lt. Commander Mario Runco, Jr. and Army Lt. Colonel James S. Voss. Mr. Thomas J. Hennen served as the payload specialist. The mission execution order (91-7) was implemented on 11 October 1991, but the scheduled launch was delayed for five days in mid-November due to a problem with the payload's IUS. A handful of optics, communications and weather instrumentation problems also cropped up during the countdown on November 24th, and the Range Safety Display System required a reload approximately half an hour before launch. Despite those problems, Atlantis' lift-off from Pad 39A went smoothly at 2344:00Z on 24 November 1991. The primary objective of the mission was to deploy a Defense Support Program (DSP) satellite approximately 6 hours and 18 minutes into the flight. The crew deployed the DSP spacecraft as scheduled at 0603Z on November 25th, but the mission was terminated three days early due to an Inertial Measurement Unit failure aboard the Shuttle. Though a landing at KSC was scheduled, Atlantis was ultimately diverted to Edwards Air Force Base for her landing. Following completion of the deorbit burn at 2131Z, Atlantis touched down on Runway 05 at 2234:42Z on 1 December 1991. The Shuttle was ferried back to KSC on 7 and 8 December 1991.[102]

Figure 95: Latest Generation Defense Support Program Satellite 1991

The Defense Department continued to support heavy military space operations at the Cape in the 1990s, but some of its most remarkable military payloads were boosted into space by light or medium launch vehicles. The first light/medium military space operations were conducted at the Cape before the end of 1959. In the decades that followed, some vehicles were launched under purely Air Force/contractor auspices. Others were contracted through NASA. In the next chapter, we will look at medium and light military space operations and the programs they supported after 1970. In the final chapter, we will review studies and proposals for future launch vehicles and facilities to support a whole new generation of military space operations at the Cape.[103]

Chapter Two Footnotes

6555th Aerospace Test Group
The Commander's office and most of the Support Division's people were located at Patrick Air Force Base. The ATLAS Systems Division and the TITAN III Systems Division operated out at the Cape's launch facilities approximately 20 miles north of Patrick.

Space Launch Vehicle Systems Division
The new division was created by joining the ATLAS Mission Management Branch and ATLAS/ AGENA Launch Operations Branch with the TITAN III Systems Division's Engineering and Launch Operations Branch, the Launch Services Branch, and the Program Management Branch.

Satellite Systems Division
The Satellite Systems Division had no formal organizational structure per se, but personnel assigned to various payload programs (e.g., Defense Support Program, DSCS, FLTSATCOM, GPS, etc.) reported directly to the Division Chief.

personnel were transferred
The 1st Space Launch Squadron was created from the 6555th's DELTA II program resources previously assigned to the Medium Launch Vehicle Division. The 1st was assigned directly to ESMC, and its first commander was Lt. Colonel R. M. Moyer. The Squadron was authorized 16 officers, 12 enlisted people and five civilians by the end of December 1991. The remaining portions of the Medium Launch Vehicle Division became the ATLAS II CTF. The ATLAS II CTF was directed by Lt. Colonel J. T. Brock, and it was authorized 11 officers, seven enlisted people and four civilians at the end of October 1991. The Space Launch Vehicle Systems Division became the TITAN IV CTF, and it was directed by Lt. Colonel William H. Barnett. By the end of October 1991, the TITAN IV CTF was authorized approximately 48 officers, 32 enlisted people and 13 civilians. The Spacecraft Division and the Resource Management Division (a.k.a. Programs/ Analysis Division) continued to operate after the transfer to AFSPACECOM as "Payload Operations" and "Ops Resource Management" under Colonel Michael R. Spence in his capacity as ESMC's Deputy for Launch Operations. Payload Operations was authorized 23 officers, 27 enlisted people and seven civilians in October 1991. During the same period, Ops Resource Management had authorizations for three officers, four enlisted people and six civilians.

45th Spacecraft Operations Squadron
The 45th Spacecraft Operations Squadron's first commander was Lt. Colonel Ivory J. Morris. The Squadron was authorized 24 officers, 23 enlisted people and seven civilians by the middle of January 1992.

launch and spacecraft agency training aids
Training lesson plans existed in the 6555th as far back as the old LARK and MATADOR programs in the early 1950s, but much of the 6555th's training program under AFSC was oriented toward apprenticeship (i.e., training someone by having that person watch another person perform tasks). Our knowledge of Test Group training is thus largely undocumented. In the interest of making their new launch resources "operational," AFSPACECOM insisted on written study guides and operating instructions at the squadron level (i.e., like other operational commands). Following ESMC's transfer to AFSPACECOM, the 1st Space Launch Squadron, the CTFs and the Payload Operations Office began writing the required study guides and operating instructions.

The Launch Controller, Anomaly Team Chief and Launch Weather Officer
The Weather Officer was assigned to the 45th Weather Squadron, but the other officials came from the 1st Space Launch Squadron.

updates from the SLOCs
On launch day, four SLOCs were assigned to supervise the following operations: 1) white room closeout, tower walkdown and Mobile Service Tower (MST) rollback, 2) lanyard tensioning, Umbilical Tower walkdown and MST rollback, 3) launch deck closeout, solid motor arming and MST rollback, and 4) postlaunch securing and scrub turnaround tasks. In addition to this supervision, the Squadron's Chief of Operations Flight conducted his own walkdown of both towers to ensure equipment was properly configured for the MST rollback.

Johnson Controls World Services
Johnson Controls acquired the old range support contractor, Pan American World Services, in February 1989. The company kept the Pan Am logo until July 1990. Pan American World Services had supported the Eastern Range (a.k.a., Eastern Test Range) under an unbroken string of contracts since December 1953.

Aerospace Systems Division and the Space Launch Systems Division
On 12 June 1990, the ESMC Vice Commander approved the merger of those two divisions into a new 37-member Heavy Launch Vehicle Division. Nine personnel were transferred from the Aerospace Systems Division to the Director of Quality's office and other divisions.

MLV and Payloads Division
As the name suggests, the MLV and Payloads Division was also responsible for ATLAS and DELTA medium launch vehicle (MLV) quality assurance.

executive agent for the Space and Missile Systems Center
Systems Program Offices (SPOs) assigned to the Space and Missile Systems Center "owned" the various spacecraft contracts and exercised authority over the individual Payload Support Contractors (PSCs), but the Field Program Manager (FPM) managed individual programs at the Cape and KSC on behalf of the various SPOs. The reader should note that this relationship went

back to the 6555th Aerospace Test Group, the Space Systems Division and the Division's predecessor organizations. Space Systems Division became the Space and Missile Systems Center as a result of the merger of AFSC and AFLC into Air Force Materiel Command on 1 July 1992. The Space and Missile Systems Center was headquartered in Los Angeles, California.

one to two years before a spacecraft arrived
Typically, detailed spacecraft planning and scheduling began 18 months before launch, but processing requirements might be considered as much as four years before launch.

Chief of the 6555th Test Group's TITAN III Systems Division
Lt. Colonel Ansel L. Wood became "Acting Commander" of the Division after Lt. Colonel Conti transferred to a new assignment at Tinker Air Force Base, Oklahoma in November 1971. Lt. Colonel Robert D. Woodward succeeded Lt. Colonel Wood as TITAN III Systems Division Chief in May 1972. Lt. Colonel Edwin W. Brenner became Chief of the TITAN III Systems Division after Colonel Woodward assumed command of the 6555th Aerospace Test Group on 16 July 1973. Woodward's predecessor, Colonel Arthur W. Banister, had succeeded Colonel Davis P. Parrish as Group Commander on 24 August 1972.

Integrate-Transfer-Launch (ITL) Area
The ITL consisted of launch complexes 40 and 41, the Vertical Integration Building (VIB), the Solid Motor Assembly Building (SMAB), the Solid Rocket Motor Processing Area, offices, warehouse space, railroads and the TITAN IIIC transporter system. The Solid Rocket Motor Processing Area consisted of a Missile Inert Storage Building for warehousing and processing solid rocket motor components, Segment Arrival Storage (SAS) areas, and a Receiving Inspection Storage (RIS) Building.

Hangar L and Missile Assembly Buildings I and II
All payload fairing activities at Hangar L were phased out in July 1971, and the building was returned to the Air Force Eastern Test Range (AFETR) organization. TITAN payload fairing storage and processing operations shifted to missile assembly buildings I and II following the completion of modifications to those buildings in July and August 1971. Missile Assembly Building I (MAB I) was used for payload fairing processing operations (e.g., the application of thermal protective coatings, ordnance installation, mechanical and electrical testings and fairing modifications). Missile Assembly Building II provided an air-conditioned storage area for flight-ready payload fairings.

companies' employees
On 1 July 1971, Martin Marietta had 232 people working its TITAN III operations at the Cape. During the same period, United Technologies had 49 people on the Range, and DELCO and Aerojet had 14 employees apiece. Aerospace Corporation was represented on the Eastern Test Range with 19 employees. By January 1973, TITAN III Systems Division authorizations had declined to 18 officers, 31 airmen and 12 civilians and actual manning dropped to around 60 people. The ratio of contractors to Division personnel changed to around 6:1 at that time, but it gradually moved back to around 4:1 in later years.

Defense Support Program (DSP)
According to a synopsis of the DSP program printed in Airman Magazine in September 1993 (Volume XXXVIII, Number 9), Defense Support Program payloads were designed to detect missile launches, space launches and nuclear detonations. From the early 1970s onward, the DSP system provided an uninterrupted early warning capability, feeding data to NORAD and (later) U. S. Space Command early warning centers at Cheyenne Mountain, Colorado. The DSP constellation of satellites orbited approximately 22,000 miles above the Earth and employed infrared sensors to detect missile and space booster exhaust plumes against Earth's background. Though the first-generation DSP satellites weighed approximately 1,800 pounds apiece, the latest DSP spacecraft weighed about 5,000 pounds. The newest DSP satellites measured 32 feet 8 inches long and 22 feet in diameter. TRW was the prime contractor for DSP spacecraft.

launched the first two Phase II Defense Satellite Communications Program (DSCP) satellites
The flight plan called for the injection of the TITAN IIIC's second stage, transtage and payload into an initial 82 x 108-nautical-mile parking orbit. The transtage and payload proceeded to a highly elliptical transfer orbit (e.g., 19,571 nautical miles at its apogee) and a final near synchronous orbit (19,494 x 19,323 nautical miles) before each of the two 1,130-pound satellites were injected into their proper orbits. One of the satellites was placed above the Atlantic Ocean, and the other was positioned over the Pacific. The Air Force Satellite Control Facility (AFSCF) in Sunnyvale, California encountered some command problems with the spacecraft initially, but both satellites were soon functioning normally. The satellites were built by TRW to meet the military's demand for high capacity, super-high frequency secure voice and data communications. Each DSCP satellite could relay as many as 1,300 duplex voice channels in an anti-jamming, secure voice environment.

TITAN IIIC vehicles (e.g., C-25 and C-28)
Core vehicles C-28 and C-25 arrived at the Cape on 25 October 1973 and 17 April 1974 respectively. Following their Acceptance CSTs, Launch Vehicle C-28 was accepted by the Air Force on 18 April 1974, and Launch Vehicle C-25 was accepted on 27 September 1974. The next core vehicle-C-29-arrived at the Cape on 24 October 1974. Erection of that vehicle began in the VIB on 28 October 1974, after C-28 was transferred to a storage stand in Cell #4 of the VIB. In effect, C-29 bumped C-28 in the launch lineup.

prelaunch testing after the lightning strikes
In addition to that testing, Stage III propellants and fuel were off-loaded, and all flight ordnance electrical systems were disconnected and checked for damage.

command of the 6555th Aerospace Test Group
During the three-week interval between the two commanders, the 6595th Aerospace Test Wing Commander (Colonel William C. Chambers) visited the Cape and assumed temporary command of the 6555th Aerospace Test Group.

Division was merged
The TITAN III Systems Division had 16 officers, 22 airmen and 13 civilians assigned to its various activities during the summer of 1975, but it lost Major Jerry H. Freer and his Space Satellite Systems branch to the new Satellite Systems Division on 1 November 1975. (The Satellite Systems Division also took over the Satellite Assembly Building in addition to Building 34705, Complex 14's ready building and the north half of Hangar AA.) What was lost in satellite systems was more than gained back in ATLAS booster facilities and people under the new division: the new Space Launch Vehicle Systems Division was authorized 20 officers, 31 airmen and 13 civilians at the end of 1975, and it had 19 officers, 30 airmen and 13 civilians assigned to its activities during that period. It also picked up ATLAS/AGENA facilities on Complex 13.

Launch Vehicle C-30
Launch Vehicle C-30 was mated to its SOLRAD and LES payloads on February 18th and February 25th, and the Launch CST was completed on 7 March 1976. The countdown for the mission got underway at Complex 40 on March 14th. Vehicle lift-off occurred at 2025:39 Eastern Standard Time on the 14th, which was equivalent to 0125:39 Greenwich Mean Time on 15 March 1976.

Lincoln Experimental Satellites
The Lincoln Experimental Satellites were experimental communications satellites built for the Air Force by the Massachusetts Institute of Technology. Each LES was ten feet long and weighed approximately 1,000 pounds. The satellites were placed in synchronous Earth orbit at an altitude of 22,300 miles to experiment with improved methods for maintaining voice or digital data circuits among widely separated, fixed or mobile communications terminals. The circuits were jam-resistant to allow operation in a hostile environment.

SOLRAD 11A and 11B
The SOLRAD spacecraft were built by the Naval Research Laboratory to monitor solar conditions and forecast disturbances severe enough to affect long-range communications and navigation systems. Each SOLRAD was 15 inches in diameter and 58 inches high, and each weighed about 400 pounds.

Eastern Daylight Time
During the 1970s and most of the 1980s, Eastern Daylight Time was used from 2 a.m. on the last Sunday in April to 2 a.m. on the last Sunday in October. During that portion of the year, clocks were advanced one hour ahead of standard time. Thus, Eastern Daylight Time was one hour later than Eastern Standard Time and four hours earlier than Greenwich Mean Time. On 8 July 1986, President Reagan signed a bill to move the start of Daylight Savings Time to the first Sunday in April.

Remote Vehicle Checkout Facility (RVCF)
The RVCF was an extension of the New Hampshire Tracking Station, and it was used to determine if commands sent from a remote tracking facility (in this instance, the New Hampshire Tracking

Station) could be received by a satellite in the RVCF. Other tests in the RVCF also confirmed a satellite's ability to send telemetry to a remote tracking facility. The 6555th's Satellite Systems Division was responsible for the RVCF.

C-31

As we noted earlier, Core Vehicle C-31 was placed in storage in Cell #3 of the VIB on 6 December 1976. It was moved to Cell #4 and erected on Transporter #3 on 2 June 1977, and then it was moved to the SMAB on June 3rd. The core vehicle and solid rocket motors were mated a few days later, and Launch Vehicle C-31 was moved out to Complex 40 on June 16th. The vehicle remained at Complex 40 for ten weeks to support a scheduled mission, which, unfortunately, did not materialize. The vehicle was returned to the SMAB on 6 September 1977, and the solid rockets were demated from the core. Core Vehicle C-31 returned to Cell #3 for storage on September 19th. The core was retested in Cell #2 toward the end of September, and it was put back in storage in Cell #3 on 8 October 1977. Interestingly enough, the vehicle's transtage (Stage III) had been used for a shock test at the Arnold Engineering Center in Tennessee, but it had been refurbished for flight prior to its arrival at the Cape in June 1976. The transtage had been flight-worthy in the summer of 1977, but the extended wait in storage began to take its toll: a crack in one of the transtage's longerons was noticed on 21 February 1978, and soon more cracks and soft aluminum patches were discovered. A decision was made to replace the transtage's control module entirely, and the transtage was shipped back to Martin Marietta's plant in Denver on 29 August 1978 to have the work done. The transtage was repaired, and it was returned to the Cape on 8 January 1979. Core Vehicle C-31 was moved out of storage on 16 January 1979, and it was transferred to Cell #2 of the VIB for reacceptance testing. The Reacceptance CST was completed successfully on March 9th, and the core vehicle was moved to the SMAB for its solid rocket mate on 14 March 1979.

Launch Vehicle C-37

The Space Division history for FY 1980 notes only two significant problems with C-37: 1) replacement of the vehicle's Stage I hydraulic pump during acceptance testing and 2) a battery malfunction five days before launch.

unscheduled five-minute hold

The hold was called to allow the contractor time to complete his electrical instrumentation checkout of the launch vehicle's third stage.

IUS Pathfinder Test Vehicle (PTV-C)

The PTV-C had arrived at the SMAB on 21 July 1981, and it completed its processing there on 7 January 1982.

operational IUS

The first operational IUS (IUS-1) was assigned to NASA's Tracking and Data Relay System satellite (TDRS-A), which was scheduled for launch on Challenger's maiden flight (i.e., the sixth Shuttle flight in the program). IUS-2 was launched on 30 October 1982, and IUS-1 was launched later, on 4 April 1983. As a point of interest, IUS-1 completed its acceptance testing on 22 May

1982, and it was moved to the SMAB on July 19th. After processing was completed at the SMAB, IUS-1 was moved to NASA's Vertical Processing Facility (VPF) on Merritt Island on November 8th. The IUS was mated to the TDRS-A satellite at the VPF on 24 November 1982. The IUS and spacecraft were installed in the orbiter's cargo bay on 24 February 1983. Following Challenger's launch on 4 April 1983, IUS-1 and the TDRS-A satellite were spring-ejected from the orbiter's cargo bay about ten hours into the mission. Though the IUS' first stage "burn" went well, the second stage's rocket nozzle malfunctioned during the second stage burn. The Air Force Satellite Control Facility in California and a ground control station at White Sands, New Mexico managed to separate the IUS from the spacecraft, but the second stage burn was cut off early. Several months of altitude correction were required before the TDRS-A reached the proper geosynchronous orbit and entered normal service.

DSCS II/III payload
The payload consisted of two DSCS satellites: one operational DSCS II communications satellite and the first flight model DSCS III satellite. (Author's note: a contract for full-scale development of the DSCS III had been awarded to General Electric in 1977. Under that contract, GE agreed to build and test one qualification model of the DSCS III satellite and deliver two flight models of the satellite. The company also agreed to build two engineering models of the electronic equipment, computer programs and communications hardware that would be needed to control the satellites from the ground.) The operational DSCS II satellite was nine feet in diameter and nine feet long. It weighed approximately 1,350 pounds. The DSCS III satellite measured 6 x 6 x 7 feet. In its operational configuration, the DSCS III weighed approximately 2,500 pounds.

Greenwich Mean Time
Mission-related times were presented in Greenwich Mean Time (GMT) in 1982 and later-year ESMC and 45th Space Wing histories to avoid the confusion inherent in Eastern Daylight Time and Eastern Standard Time references. The Air Force specified Greenwich Mean Time with a "Z" (Zulu) suffix. The "Z" designation will be used for all times appearing in the remainder of this chapter.

transtage
The TITAN 34D Program Requirements Document was revised in early 1982 to allow the transition of classified payloads from TITAN IIICs to TITAN 34Ds equipped with transtages (TSs) as well as IUSs. The transtages were required due to slips in the IUS program schedule.

launch operations ground to a halt
As we noted in Chapter I, the TITAN 34D-9 launch disaster at Vandenberg AFB on 18 April 1986 effectively grounded TITAN 34D launch operations on both coasts until an aggressive recovery program was implemented to reconfirm the TITAN 34D's reliability.

D-13's acceptance testing
Core Vehicle D-13 was moved from Cell #1 to Cell #2 after Core Vehicle D-11 was moved from the VIB to the SMAB in February 1984. D-13's first stage was replaced in Cell #2, and the vehicle was moved back to Cell #1 for acceptance testing.

Under the command
Colonel Martinelli assumed command of the 6555th on 25 February 1985. His predecessor, Colonel Charles A. Kuhlman, commanded the Test Group for one of the longest periods on record (10 August 1981 through 24 February 1985). Colonel Kuhlman's predecessor was Colonel Walter S. Yager, who succeeded Colonel George L. Rosenhauer as Test Group Commander on 28 February 1980.

NDT facility
The facility consisted of three cells-A, B and C. Cells A and B were identical 20 x 32-foot reinforced concrete rooms with removable roof panels which allowed rocket motors to be lowered into the cells. Cell C was built the same way, but its floorspace was considerably larger (e.g., 47 x 58 feet). A central control area joined cells A and B in one wing of the facility with Cell C in the other wing. Together with its environmental enclosure and 85-ton crane, the facility cost approximately $27.8 million to build and equip.

refurbish Launch Complex 41
Complex 41 had been deactivated at the end of 1977 after NASA's last VOYAGER mission was launched to the outer planets. For the next eight years, the Cape's corrosive salt air environment ate away at the site unchecked. Under the circumstances, Complex 41's renovation had to be extensive. The project included: 1) tearing out or modifying structural, mechanical and electrical systems, 2) sandblasting, priming and painting Complex 41's towers, 3) changing fuel systems and 4) installing air pollution control devices on fuel and oxidizer systems. Though Martin Marietta was responsible for the project, the Eastern Space and Missile Center formed a TITAN IV Site Activation Working Group (SAWG) to monitor the project's progress. The Test Group and various ESMC agencies (e. g., contracts, quality assurance, range facilities engineering, range systems development, safety, range contractors, ETR program management and the Aerospace Corporation) sent representatives to the SAWG's monthly meetings.

corrosion was discovered
In the fall of 1987, signs of corrosion were detected under the UT's new paint job. Some corrosion was also discovered on a propellant piping chase in May 1988, but a preliminary propellant "hot flow" review in September 1988 confirmed the integrity of Complex 41's fueling system. After the first TITAN IV launch in June 1989, it was apparent that some portions of the towers would have to be sandblasted, primed and painted all over again. At the end of July 1989, the Test Group asked Range Facilities Engineering to task the Launch Base Services contractor to reaccomplish the work. Pan American's remedial work was accomplished in the summer of 1990.

CENTAUR Payload Operations Control Center (CPOCC)
The CPOCC was originally designed to control and monitor CENTAUR vehicle, spacecraft and support systems operations for Shuttle/CENTAUR missions. The facility was linked to spacecraft/ CENTAUR integration at the SPIF and Shuttle/CENTAUR systems at NASA's Vertical Processing

Facility on Merritt Island. It was also linked to ATLAS/CENTAUR integration and testing systems on Launch Complex 36.

second floor redesigned
Sterns-Roger was contracted to redesign and modify the second floor to provide Test Group and higher headquarters commanders and managers with information displays and communications outlets to their agencies. The modifications included new walls, LMCC consoles, fire suppression and air-conditioning systems, new cable raceways and consoles for the Command Management Control Center (CMCC). The work was completed in December 1988. The LMCC was activated in July 1989, and it was fully operational in the spring of 1990.

three alternatives
The three alternatives were: 1) modify the existing SMAB, 2) build a duplicate of the old SMAB, and 3) build an improved SMAB (i.e., the SMARF). A new improved SMAB would cost more to build than modifications to the old SMAB, but (surprisingly) it would be cheaper to build than a duplicate of the old SMAB. A persuasive point in favor of the SMARF was that it could be designed to allow the stacking and storage of two flight sets of SRMUs in a vertical, checked out configuration, thereby optimizing the operators' ability to return a TITAN IV to the SMARF if a problem developed at the launch pad. (Author's note: though the basic TITAN IV vehicle was designed to carry two seven-segment solid rocket motors, the new three-segment SRMU design was expected to be introduced once TITAN IV operations were underway. Each of the SRMU segments would measure 12 x 30 feet, as opposed to the seven 10 x 10-foot motor segments carried on the basic TITAN IV vehicle. Consequently, an SRMU stack was significantly taller and heavier than a basic seven-segment SRM stack, and it posed a bigger storage problem. Both types of solid rocket motors would continue to be used with the TITAN IVs, but the SRMUs would be used for heavier payloads and/or higher energy orbits.)

contract
Modifications to the contract raised its value to more than $42 million before the SMARF was completed in October 1991.

launch vehicle (D-8)
Core Vehicle D-8 arrived at the Cape on 19 February 1985. It was erected on Transporter #2 in Cell #1 of the VIB by February 27th, and its Acceptance CST was completed there on 7 May 1985. The vehicle was transported to the SMAB for solid rocket mating on July 15th. As luck would have it, a severe thunderstorm developed near the SMAB on July 16th, and a lightning strike evidently damaged some of D-8's guidance system components. The damage was discovered after Launch Vehicle D-8 was moved to Complex 40, but Baseline CSTs at the pad were completed successfully at the end of August 1985 after the components were repaired or replaced. In the meantime, news of the TITAN 34D-7 failure at Vandenberg prompted a complete reassessment of D-8's propulsion systems, and that new requirement delayed the launch by at least several months. A second Baseline CST was completed on D-8 on 28 February 1986, but the vehicle had to be returned to the VIB for a thorough solid rocket motor inspection in July 1986 after the TITAN 34D-9 launch

failure in April grounded the TITAN fleet. A Reassessment CST was completed on D-8 in the VIB on 27 May 1987, and the core vehicle was moved to the SMAB for another solid rocket mate on July 8th. Launch Vehicle D-8 was moved out to Complex 40 on July 10th, and a third Baseline CST was required as a retest for extensive "black box" (avionics) replacements and other improvements. Final preparations for the November 1987 launch continued in a relatively routine fashion through the summer and early fall of 1987.

unscheduled holds
The first unscheduled hold occurred at T minus 10 minutes when vessels were detected in the launch danger zone. A 29-minute delay in the countdown was required to get the intruders to safety. The second unscheduled hold was called at T minus five minutes when the umbilical retract mechanism on Solid Rocket Motor #1 failed to lock. A technician was sent out to the pad to lock the mechanism manually, and the hold was terminated after 69 minutes.

beginning of the TITAN IV era
The first TITAN IV liquid rocket engines arrived at the Cape on 18 December 1987, and the roll-out ceremony for stages I and II of the TITAN IV pathfinder vehicle took place at Complex 41 on 14 January 1988. Colonel Bourne succeeded Colonel Dominick R. Martinelli as Test Group Commander on 15 January 1988. As noted earlier, Colonel Michael R. Spence assumed command of the 6555th Aerospace Test Group on 2 October 1990, and he succeeded Colonel Bourne.

countdown
Only one nine-minute unscheduled hold was required at L minus five minutes to clear the launch danger area. The countdown proceeded smoothly thereafter.

ATLAS I/CENTAUR upper stage flight failures
Both of the failures involved commercial ATLAS I/CENTAUR vehicles launched from Launch Complex 36B on the Cape. The first failure occurred on 18 April 1991 when one of the CENTAUR's Pratt & Whitney engines failed to start about six minutes after lift-off. Range Safety sent destruct commands, and the vehicle's destruction was confirmed seconds later. The second failure occurred on 22 August 1992. The flight on that date went well until the CENTAUR upper stage malfunctioned, and Flight Control Officers (FCOs) had to destroy the vehicle about eight minutes after launch.

Lt. Colonel George L. Rosenhauer
Lt. Colonel Rosenhauer assumed the position of STS Division Chief on 4 August 1975.

Chief of the STS Division
Lt. Colonel Charles A. Kuhlman assumed the post of Assistant Chief, STS Division on 10 July 1978. He succeeded Lt. Colonel Green as STS Division Chief in 1979.

transferred
The transfer raised the Division's assigned strength to ten officers and one civilian on 1 October 1977. Eleven officers and one civilian were assigned to the Division by the end of 1977.

IUS was mission-specific
The Space and Missile Systems Center (SMC) in Los Angeles provided overall program management for the IUS, and Boeing Aerospace was SMC's prime contractor for IUS operations. During the period in question, the 6555th Aerospace Test Group served as the "on-scene" launch support agency for Shuttle/IUS operations at KSC and the Cape. The 45th Operations Group's Launch Vehicle Directorate also contributed to the effort after the 45th Space Wing was activated on 12 November 1991.

support an astronaut bailout
One HC-130 aircraft was on cockpit alert at Patrick. One KC-130 was on ready alert at Cherry Point, North Carolina, and two E-2C Hawkeyes were on ready alert at the Naval Air Station in Norfolk, Virginia. One Coast Guard cutter (with helicopter) was stationed about 100 nautical miles downrange, and one Navy ship (with helicopter) was positioned 150 nautical miles downrange. An HC-130 aircraft orbited about 175 nautical miles downrange, and C-130s were on cockpit alert at Moron Air Base (Spain), Banjul (the Gambia), Zaragoza (Spain) and Ben Guerir (Morocco). Two P-3C aircraft were also on ready alert at Jacksonville, Florida and at Lajes in the Azores. Two H-60 Army medical evacuation helicopters and four H-1 support helicopters were also on alert at Edwards Air Force Base, California to handle medical, security and photographic support tasks.

orbiter's three main engines
The solids produced 2,650,000 pounds of thrust apiece, and each of the orbiter's main engines produced 390,000 pounds of thrust at sea level. At altitude, each main engine's performance increased to approximately 488,000 pounds of thrust.

SYNCOM IV
As we noted earlier, the SYNCOM IV constellation was sent into space to replace aging FLTSATCOM satellites that provided worldwide defense communications for submarines and ships at sea. The SYNCOM IV was a Hughes HS381 satellite with a life expectancy of ten years. In its stowed configuration, the satellite measured 14 feet 1 inch by 13 feet 9 inches. It weighed 17,081 pounds. The satellite was deployed horizontally by means of a rotational pivot. The SYNCOM IV was pushed off the port side of the Shuttle's cargo bay at a velocity of 1.4 feet per second. The satellite's perigee kick motor (e.g., a MINUTEMAN III TU-882 upper stage) fired 45 minutes later, injecting the payload into an elliptical transfer orbit. Subsequently, a liquid apogee motor circularized the SYNCOM IV's orbit at an altitude of 19,300 nautical miles and established final synchronous orbit.

all-military Shuttle mission
The term "all-military Shuttle mission" refers only to the major payload or payloads carried into space, not the composition of the crew. Civilians were not excluded from crew duty, and a civilian crewmember (Dr. Kathryn C. Thornton) served on an all-military mission in November 1989.

two unscheduled holds
The first hold was called at 1255Z after a cargo vessel was detected in the surveillance area. The second unscheduled hold was called at 1305Z for foul weather, and it lasted 45 minutes.

Shuttle mission at 0750:22Z on 28 February 1990
The mission experienced a total of five launch delays and scrubs between February 21st and February 28th. The launch was slipped 24 hours due to a crew member illness on February 21st. A forecast of poor weather (compounded by uncertainty concerning the crew member's health) pushed the launch to February 24th. The countdown on the 24th was scrubbed at L minus 15 hours due to weather constraints, and another launch attempt on February 25th was scrubbed at T minus 31 seconds when the Eastern Range's Cyber B computer failed. The countdown on February 26th was scrubbed due to high upper level winds, but the final countdown on February 28th was successful despite unscheduled holds for weather in the launch area.

Shuttle Pallet Satellite II
The Shuttle Pallet Satellite II (SPAS II) was an advanced version of the SPAS I spacecraft that flew on two civilian Shuttle missions in June 1983 and April 1984. The Shuttle's Remote Manipulator System (RMS) deployed the satellite. Once the SPAS II's altitude control system was checked out, Discovery maneuvered about ten kilometers away from the spacecraft. At that distance, the SPAS II took "far-field" measurements of the Shuttle's OMS and primary control system exhaust plumes to characterize their spectral and radiometric signatures. Similar observations and measurements were made of Earth and Earthlimb fields to provide a background model for boost detection and target tracking purposes. Following the far-field measurements, Discovery moved to a distance of about 2.25 kilometers to let the SPAS II capture near-field measurements. A rendezvous followed, and the RMS grappled the SPAS II and brought it aboard. The three Chemical Release Observation (CRO) sub-satellites were also carried aboard Discovery in get-away special canisters. The CRO sub-satellites were ejected into space, and they were commanded to release three different rocket fuel clouds (e.g., monomethyl hydrazine, unsymmetrical dymethylhydrazine and nitrogen tetroxide) for observation by the IBSS and other instruments during the mission.

end of 1959
The Air Force Ballistic Missile Division supported a total of ten Air Force-sponsored THOR-ABLE, THOR-ABLE I and THOR-ABLE II space launches from Pad 17A before the end of 1959.

Chapter Two Endnotes

1. 6555th Aerospace Test Group History, 1 January - 30 June 1971, Chapter II, "Personnel;" 6555th Aerospace Test Group History, 1 January - 30 June 1970, Chapter I, "Organization and Mission;" 6555th Aerospace Test Wing History, 1 January - 30 June 1968, Chapter I, "Organization and Mission;" 6555th Aerospace Test Group History, 1 January - 30 June 1973, Supporting Document, "Concepts of 6555th Aerospace Test Group Launch Operations."

2. 6555th Aerospace Test Group History, 1 July - 31 December 1974, STS Division Historical Section "Introduction" and "Strength Resume".

3. 6555th Aerospace Test Group History, 1 January - 30 June 1975, TITAN III Systems Division Historical Section, "Mission" and "Organizational Structure;" 6555th Aerospace Test Group History, 1 July - 31 December 1975, Chapter I, "Organization" and Space Launch Vehicle Systems Division Historical Section, "Organization and Mission;" 6555th Aerospace Test Group History, 1 January - 30 June 1976, Satellite Systems Division Historical Section, "Organization and Mission;" 6555th Aerospace Test Group History, 1 July - 31 December 1977, Space Launch Vehicle Systems Division Historical Section, "Introduction;" 6555th Aerospace Test Group History, January - September 1978, pp. 21, 23; 6555th Aerospace Test Group History, October 1978 - September 1979, Volume I, pp.1, 6, 22; ESMC History, 1 October 1987 - 30 September 1988, Volume I, p. 32.

4. 6555th Aerospace Test Group History, October 1978 - September 1979, Volume I, pp. 1, 2, 3; ESMC History, 1 October 1979 - 30 September 1981, Volume I, p. 22; ESMC Organizational Chart, April 1981; ESMC History, 1 October 1982 - 30 September 1984, Volume I, pp. 3, 4; ESMC History, 1 October 1987 - 30 September 1988, Volume I, p. 33.

5. ESMC History, 1 October 1987 - 30 September 1988, Volume I, pp. 32, 33, 34.

6. ESMC/45 SPW History, 1 October 1990 - 31 December 1991, Volume I, pp. 39, 40, 42, 43; USAF Biography, "Colonel Michael R. Spence," a/o April 1992.

7. ESMC/45 SPW History, 1 October 1990 - 31 December 1991, Volume I, pp. 43, 44; 45 SPW History, 1 January - 31 December 1992, Volume I, p.15; Item, "Group deactivates after merger," *The Missileer*, 10 July 1992.

8. Interview, Mark C. Cleary with Brigadier General Jimmey R. Morrell, 45 SPW Commander, 9

March 1993; Long Range Proving Ground Division History, 1 July - 31 December 1950, pp. 171, 172, 173.

9. Squadron Operating Instruction (SOI) 55-2, 1 SLS, 22 March 1991, pp. 1, 2; SOI 55-21, 3 SLS, 3 September 1992, pp. 1, 2.

10. SOI 55-1, 1 SLS, 1 October 1991, pp. 3, 49, 50, 51; SOI 55-8, 1 SLS, 4 November 1991, p. 43; SOI 55-22, 3 SLS, 13 November 1992, pp. 1, 2, 3.

11. SOI 55-8, 1 SLS, 4 November 1991, pp. 1, 2, 3.

12. SOI 55-8, 1 SLS, 4 November 1991, pp. 9, 10.

13. SOI 55-8, 1 SLS, 4 November 1991, pp. 16, 21.

14. SOI 55-8, 1 SLS, 4 November 1991, p. 26.

15. SOI 55-8, 1 SLS, 4 November 1991, pp. 31 and 37.

16. SOI 55-8, 1 SLS, 4 November 1991, p. 49.

17. SOI 55-8, 1 SLS, 4 November 1991, p. 43.

18. SOI 55-23, 3 SLS, 16 November 1992; 3 SLS, "ATLAS III Launch Operations," 13 May 1992; 3 SLS, "Operations Control," 19 May 1992.

19. TITAN CTF Self-Study Guide (SSG) I-00, 45 OPG/LVOT, 8 July 1992, pp. 4, 5.

20. TITAN CTF SSG I-00, 45 OPG/LVOT, 8 July 1992, pp. 5, 6.

21. TITAN CTF SSG I-00, 45 OPG/LVOT, 8 July 1992, pp. 6, 7.

22. TITAN CTF SSG I-00, 45 OPG/LVOT, 8 July 1992, p. 6.

23. TITAN CTF SSG V-001, 45 OPG/LVOT, 22 June 1992, p. 2.

24. TITAN CTF SSG V-001, 45 OPG/LVOT, 22 June 1992, p. 5.

25. TITAN CTF SSG V-001, 45 OPG/LVOT, 22 June 1992, p. 8.

26. TITAN CTF SSG V-001, 45 OPG/LVOT, 22 June 1992, p. 9.

27. ESMC History, 1 October 1984 - 30 September 1986, Volume I, p. 19.

28. ESMC History, 1 October 1989 - 30 September 1990, Volume I, pp. 24, 25, 26, 27, 28; 45 SPW History, 1 January - 31 December 1992, Volume I, pp. 21, 22, 23, 24.

29. Interview, Mark C. Cleary with Lt. Colonel Ivory J. Morris, Commander, 45 SPOS, 19 November 1992; Slide Briefing, 45 SPOS, "Spacecraft, Pathway to the Future," o/a November 1992.

30. Slide Briefing, 45 SPOS, "Spacecraft, Pathway to the Future," o/a November 1992.

31. Interview, Mark C. Cleary, with Lt. Colonel Ivory J. Morris, 45 SPOS Commander, 19 November 1992; SOI 55-33, 45 SPOS, 16 December 1991; SSG A-15, 45 SPOS/DOT, "Review Operations Documents," 20 August 1992; SSG A-3, 45 SPOS/DOT, "Control Launch Base Operations," 14 July 1992, pp. 1, 2, 3.

32. SSG A-3, 45 SPOS/DOT, "Control Launch Base Operations," 14 July 1992, pp. 1, 2, 3.

33. *Ibid.*

34. SSG A-3, 45 SPOS/DOT, "Control Launch Base Operations," 14 July 1992, pp. 3, 4, 5; Summary, Major K. W. Carlson, 45 SPOS, "Local Operations Requiring Controller Coverage," undated; SOI 55-21, 45 SPOS, "Payload Operations Control," 16 December 1991; SSG A-1, 45 SPOS/DOT, 7 July 1992, pp. 11, 12; SOI 55-31, 45 SPOS, "Personnel Safety," 16 December 1991.

35. SSG A-3, 45 SPOS/DOT, "Control Launch Base Operations," 15 July 1992, pp. 3, 4; Slide Briefing, Major K. W. Carlson, 45 SPOS, "Introduction to Spacecraft Operations and Training," undated; SSG C-42, 45 SPOS/DOT, "Perform Countdown Activities," 2 July 1992, pp. 3, 4.

36. AFETR History, FY 1971, Volume 2, pp. 291, 373, 374, 375; 6555th Aerospace Test Group History, 1 January - 30 June 1971, TITAN III Systems Division Historical Section, pp. 2, 4, 8; 6555th Aerospace Test Group History, 1 July - 31 December 1971, Chapter II, "Personnel" and TITAN III Systems Division Historical Section, pp. 4, 5, 10; 6555th Aerospace Test Group History, 1 January - 30 June 1972, TITAN III Systems Division Historical Section, p.4; 6555th Aerospace Test Group History, 1 July - 31 December 1972, Command Office Historical Section, "Organization and Mission;" 6555th Aerospace Test Group History, 1 January - 30 June 1973, TITAN III Systems Division Historical Section, pp. 4, 5; Article, "Titan Chief Named Test Group Commander," *The Missileer*, 20 July 1973; Summary, "Satellites," *Airman Magazine*, Volume XXXVIII, Number 9, September 1993, p. 56.

37. 6555th Aerospace Test Group History, 1 July - 31 December 1971, TITAN III Systems

Division Historical Section, pp. 4, 8, 9; Cleary, *The 6555th*, pp. 204, 205; AFETR History, FY 1972, Volume I Part 2, pp. 357, 359; AFETR History, FY 1975, Volume I Part 2, p. 373; 6555th Aerospace Test Group History, 1 January - 30 June 1972, TITAN III Systems Division Historical Section, pp. 9, 10; AFETR History, FY 1973, Volume I Part 2, p. 354; 6555th Aerospace Test Group History, 1 July - 31 December 1972, TITAN III Systems Division Historical Section, p. 8; 6555th Aerospace Test Group History, 1 January - 30 June 1973, TITAN III Systems Division Historical Section, pp. 6, 7, 8.

38. 6555th Aerospace Test Group History, TITAN III Systems Division Historical Section, pp. 4, 8, 9.

39. 6555th Aerospace Test Group History, 1 January - 30 June 1974, TITAN III Systems Division Historical Section, pp. 7, 9; 6555th Aerospace Test Group History, 1 July - 31 December 1973, TITAN III Systems Division Historical Section, p. 10; 6555th Aerospace Test Group History, 1 July - 31 December 1974, TITAN III Systems Division Historical Section, pp. 10 and 11; AFETR History, FY 1975, Volume I Part 2, pp. 372, 373; 6555th Aerospace Test Group History, 1 January - 30 June 1975, TITAN III Systems Division Historical Section, p. 7.

40. AFETR History, July - December 1975, Volume I, Part 2, p. 276; 6555th Aerospace Test Group History, 1 July - 31 December 1975, Space Launch Vehicle Systems Division Historical Section, pp. 8, 9.

41. 6555th Aerospace Test Group History, 1 January - 30 June 1975, Command Office, "Activities," TITAN III System Division Historical Section, pp. 4, 5, and ATLAS Systems Division Historical Section, p. 3; 6555th Aerospace Test Group History, 1 July - 31 December 1975, "Organization and Mission," and Space Launch Vehicle Systems Division Historical Section, pp. 1, 6; 6555th Aerospace Test Group History, 1 January - 30 June 1976, Satellite Systems Division Historical Section, pp. 1, 2; 6555th Aerospace Test Group History, 1 January - 30 June 1977, Command Historical Section, "Change of Command," and Space Launch Vehicle Systems Division Historical Section, p. 7.

42. AFETR History, CY 1976, Volume I Part 2, pp. 2-58, 2-59, 2-60; 6555th Aerospace Test Group History, 1 January - 30 June 1976, Space Launch Vehicle Systems Division Historical Section, pp. 7, 11, 12.

43. 6555th Aerospace Test Group History, 1 July - 31 December 1976, Space Launch Vehicle Systems Division Historical Section, p. 11; Report, 6555th Aerospace Test Group, "Launch Evaluation Report, TITAN IIIC Vehicle 23C-13 (C-31)," 1 November 1979, p. 9 (information used is totally unclassified).

44. 6555th Aerospace Test Group History, 1 January - 30 June 1977, Space Launch Vehicle Systems Division Historical Section, pp. 10, 11; Pamphlet, 6555th Aerospace Test Group, "The Air

Force TITAN III and 34D," o/a January 1980, TITAN IIIC Operational Program Section.

45. Det 1 SAMTEC History Office, "Eastern Test Range Index of Missile Launchings," CY 1977, p. 29 (information used is totally unclassified); Pamphlet, 6555th Aerospace Test Group, "The Air Force TITAN III and 34D," o/a January 1980, TITAN IIIC Operational Program Section; 6555th Aerospace Test Group History, 1 January - 30 June 1977, Space Launch Vehicle Systems Division Historical Section, pp. 11, 12.

46. 6555th Aerospace Test Group History, 1 January - 30 June 1977, Space Launch Vehicle Systems Division Historical Section, p. 12; 6555th Aerospace Test Group History, 1 July - 31 December 1977, Space Launch Vehicle Systems Division Historical Section, p. 10; Report, 6555th Aerospace Test Group, "Launch Evaluation Report, TITAN IIIC Vehicle 23C-16 (C-34)," p. 7 (information used is totally unclassified.)

47. 6555th Aerospace Test Group History, January - September 1978, pp. 19, 21, 26, 27; 6555th Aerospace Test Group History, October 1978 - September 1979, p. 13; Det 1 SAMTEC History Office, "Eastern Test Range Index of Missile Launchings," CY 1978, p. 29 (information used is totally unclassified).

48. Det 1 SAMTEC History Office, "Index of Missile Launchings," CY 1978, p. 29 (information used is totally unclassified); Pamphlet, 6555th Aerospace Test Group, "The Air Force TITAN III and 34D," o/a January 1980, TITAN IIIC Operational Program Section; 6555th Aerospace Test Group History, January - September 1978, pp. 17, 21; 6555th Aerospace Test Group History, October 1978 - September 1979, Volume I, pp. 32, 42, 44.

49. 6555th Aerospace Test Group History, October 1978 - September 1979, Volume I, pp. 52, 53; Report, 6555th Aerospace Test Group, "Launch Evaluation Report, TITAN IIIC Vehicle 23C-13 (C-31)," 1 November 1979, p. 9 (information used is totally unclassified); Pamphlet, 6555th Aerospace Test Group, "The Air Force TITAN III and 34D," o/a January 1980, TITAN IIIC Operational Program Section.

50. Report, 6555th Aerospace Test Group, "Launch Evaluation Report, TITAN IIIC Vehicle 23C-16 (C-34)," 1 April 1980, p. 7 (information used is totally unclassified); Pamphlet, 6555th Aerospace Test Group, "The Air Force TITAN III and 34D," o/a January 1980, TITAN IIIC Operational Program Section.

51. Space Division History, 1 October 1979 - 30 September 1980, Volume I, p. 71; ESMC History, 1 October 1979 - 30 September 1981, Volume I, p. 475; Report, Martin Marietta, "Vehicle 23C-19 (C-37) Test Report," December 1979, pp. 4-1, 4-6, 4-10, 4-14; Pamphlet, 6555th Aerospace Test Group, "The Air Force TITAN III and 34D," o/a January 1980, TITAN IIIC Operational Program Section.

52. ESMC History, 1 October 1979 - 30 September 1981, Volume I, p. 476 (information used is totally unclassified); ESMC History, 1 October 1981 - 30 September 1982, Volume I, pp. 170, 171, 172 (information used is totally unclassified); Cleary, *The 6555th*, pp. 199, 200; Report, Martin Marietta, "Vehicle 23C-22 (C-40) Test Report," o/a 1 May 1981, pp. 4-1, 4-3, 4-6, 4-8, 4-9, 4-13, 4-17.

53. ESMC History, 1 October 1981 - 30 September 1982, Volume I, pp. 170, 171 (information used is totally unclassified); Cleary, *The 6555th*, pp. 199, 200; Report, Martin Marietta, "Vehicle 23C-21 (C-39)," o/a 1 May 1982, pp. 4-1, 4-3, 4-4, 4-6, 4-7, 4-18.

54. ESMC History, 1 October 1981 - 30 September 1982, Volume I, pp. 171, 172 (information used is totally unclassified).

55. ESMC History, 1 October 1981 - 30 September 1982, Volume I, pp. 174, 176, 177; ESMC History, 1 October 1982 - 30 September 1984, Volume I, p. 182; Space Division History, 1 October 1981 - 30 September 1982, Volume I, pp. 81, 82, 142; Space Division History, 1 October 1982 - 30 September 1984, Volume I, pp. 75, 75, 77; 6555th Aerospace Test Group History, 1 April - 30 September 1982, Space Launch Vehicle Systems Division Historical Section, pp. 1, 2; 6555th Aerospace Test Group History, 1 October 1982 - 30 March 1983, Space Launch Vehicle Systems Division Historical Section, "IUS Operations;" Summary, "Satellites," *Airman Magazine*, Volume XXXVIII, Number 9, p. 56.

56. ESMC History, 1 October 1982 - 30 September 1984, Volume I, p. 172; Space Division History, 1 October 1982 - 30 September 1983, Volume I, p. 74; 6555th Aerospace Test Group History, 1 October 1982 - 31 March 1983, Space Launch Vehicle Systems Division Historical Section, "TITAN Operations;" Almanac, Newspaper Enterprise Association, Inc., "The World Almanac," 1968 Centennial Edition, p. 682.

57. ESMC History, 1 October 1981 - 30 September 1982, Volume I, p. 194; Report, 6555th Aerospace Test Group, "Launch Evaluation Report, Titan 34D Vehicle D-10/TS-2," 30 April 1984, pp. 8, 15, 29, 30, 31, 32, 33, 34, 35, 36, 37, 38, 39. (Information used is totally unclassified); 6555th Aerospace Test Group History, 1 October 1982 - 31 March 1983, Space Launch Vehicle Systems Division Historical Section, "TITAN Operations;" 6555th Aerospace Test Group History, 1 October 1983 - 31 March 1984, Space Launch Vehicle Systems Division, "TITAN Operations;" Message, 6555th Aerospace Test Group to HQ AFSC, "Launch of TITAN 34D Vehicle D-10," 310500Z, January 1984.

58. 6555th Aerospace Test Group History, 1 October 1982 - 31 March 1983, Space Launch Vehicle Systems Division Historical Section, "TITAN Operations;" 6555th Aerospace Test Group History, 1 October 1983 - 31 March 1984, Space Launch Vehicle Systems Division Historical Section, "TITAN Operations;" 6555th Aerospace Test Group History, 1 April - 30 September 1984, Space Launch Vehicle Systems Division Historical Section, "TITAN Operations (LVT);"

Report, 6555th Aerospace Test Group, "Launch Evaluation Report, Titan 34D Vehicle D-11/TS-2," o/a 30 July 1984, pp. 8, 28, 29, 30, 31, 32, 33, 34, 35, 36, 37, 38, 39 (information used is totally unclassified); Message, SAMTO (at CCAFS) to HQ AFSC, "Launch of TITAN 34D-11/TS-2 Vehicle from Cape Canaveral AFS," 161330Z April 1984; Report, RCA International Service Corporation, "Land-Based Radar Test Report, Test 3023," 11 May 1984, p. 1.

59. 6555th Aerospace Test Group History, 1 April - 30 September 1983, Space Launch Vehicle Systems Division Historical Section, "TITAN (LVT);" 6555th Aerospace Test Group History, 1 October 1983 - 31 March 1984, Space Launch Vehicle Systems Division Historical Section, "TITAN Operations (LVT);" 6555th Aerospace Test Group History, 1 April - 30 September 1984, Space Launch Vehicle Systems Division Historical Section, "TITAN Operations (LVT);" Report, 6555th Aerospace Test Group, "Launch Evaluation Report, TITAN 34D Vehicle D-13," 1 April 1986, pp. 14, 15, 22, 28, 29, 30, 31, 32, 33, 34, 35, 36, 37, 38, 39 (information used is totally unclassified); Letter, Eastern Test Range Directorate of Range Operations, "Post Launch Debriefing on Test 2057, TITAN 34D/TS, 21 December 1984," 4 March 1985.

60. ESMC History, 1 October 1979 - 30 September 1981, Volume I, Appendix 6, "Key Personnel," p. 658; ESMC History, 1 October 1984 -30 September 1986, Volume I, pp. 79, 441, 448; ESMC History, 1 October 1986 - 30 September 1987, Volume I, pp. 220, 221, 223, 227.

61. ESMC History, 1 October 1986 - 30 September 1987, Volume I, pp. 201, 202, 203; ESMC History, 1 October 1987 - 30 September 1988, Volume I, pp. 179, 181; Cleary, *The 6555th*, p. 207, ESMC History, 1 October 1989 - 30 September 1990, Volume I, p. 122.

62. ESMC History, 1 October 1987 - 30 September 1988, Volume I, pp. 181, 182; ESMC History, 1 October 1989 - 30 September 1990, Volume I, pp. 122, 123, 124, 125.

63. ESMC History, 1 October 1986 - 30 September 1987, Volume I, pp. 218, 219, 220; ESMC History, 1 October 1987 - 30 September 1988, Volume I, pp. 185, 186; ESMC History, 1 October 1988 - 30 September 1989, Volume I, pp. 196, 197, 198; ESMC History, 1 October 1989 - 30 September 1990, Volume I, pp. 130, 131, 132, 133, 134, 135, 136.

64. ESMC History, 1 October 1987 - 30 September 1988, Volume I, pp. 182, 183, 184; ESMC History, 1 October 1988 - 30 September 1989, Volume I, pp. 198, 199; ESMC History, 1 October 1989 - 30 September 1990, Volume I, pp. 136, 137, 138; ESMC/45 SPW History, 1 October 1990 - 31 December 1991, Volume I, pp. 180, 181, 183; 45 SPW History, 1 January - 31 December 1992, Volume I, pp. 98. 99, 100, 101.

65. ESMC History, 1 October 1987 - 30 September 1988, Volume I, pp. 286, 289, 290, 291; Report, Martin Marietta, "Vehicle 05D-4 (T34D-8) Test Report," March 1988, pp. 4-26, 5-1, 5-2, 5-3, 5-4, 5-5.

66. ESMC History, 1 October 1987 - 30 September 1988, Volume I, pp. 291, 292; Space Systems Division History, October 1988 - September 1989, Volume I, pp. 120, 121; Report, Martin Marietta, "Test Report, Vehicle 05D-5 (34D-3)," December 1988, pp. iv, v.

67. ESMC History, 1 October 1987 - 30 September 1988, Volume I, pp. 84, 187, 188, ESMC History, 1 October 1989 - 30 September 1990, Volume I, p. 68.

68. ESMC History, 1 October 1987 - 30 September 1988, Volume I, pp. 188, 189; ESMC History, 1 October 1988 - 30 September 1989, Volume I, pp. 187, 188.

69. ESMC History, 1 October 1988 - 30 September 1989, Volume I, pp. 352, 353; Space Systems Division History, October 1988 - September 1989, Volume I, p. 128.

70. Space Systems Division History, October 1988 - September 1989, Volume I, pp. 128, 129.

71. Report, Martin Marietta, "Test Report, Vehicle 05D-6 (34D-16)," July 1989, pp. v, vii, 4-1, 4-2, 4-4, 4-5, 5-1, 5-2, 5-3, 5-4, 5-5; ESMC History, 1 October 1988 - 30 September 1989, Volume I, p. 346.

72. ESMC History, 1 October 1988 - 30 September 1989, Volume I, pp. 347; Report, Martin Marietta, "Test Report, Vehicle 05D-7 (34D-2)," undated, pp. v, 4-1, 4-5, 5-1, 5-2, 5-3, 5-4.

73. ESMC History, 1 October 1989 - 30 September 1990, Volume I, pp. 300, 301; Report, Martin Marietta, "Post Launch Test Report, Vehicle K4 (45H-2/TIV-4)," 16 October 1990, pp. 1-1, 5-1, 5-2, 5-3, 5-4.

74. Space Systems Division History, October 1989 - September 1990, Volume I, p. 154 (information used is totally unclassified); ESMC/45 SPW History, 1 October 1990 - 31 December 1991, Volume I, p. 320; Report, Martin Marietta, "SR 91-7, Post Launch Test Report, TITAN IV/ IUS (K6)," 4 December 1991, pp. 1, 4, 9, 14, 19, 20, 21, 22, 23.

75. ESMC/45 SPW History, 1 October 1990 - 31 December 1991, Volume I, pp. 330, 331; 45 SPW History, 1 January - 31 December 1992, Volume I, pp. 217, 218, 226; Interview, Mark C. Cleary with Ms. Jennifer Bevins, 45 OPG/LV, 3 September 1993.

76. ESMC/45 SPW History, 1 October 1990 - 31 December 1991, Volume I, pp. 171, 174, 175, 176.

77. 45 SPW History, 1 January - 31 December 1992, Volume I, pp. 94, 96; Interview, Mark C. Cleary with Ms. Jennifer Bevins, 45 OPG/LV, 3 September 1993.

78. 6555th Aerospace Test Group History, 1 July - 31 December 1974, STS Division Historical

Section, p. 1; 6555th Aerospace Test Group History, 1 January - 30 June 1975, STS Division Historical Section, p. 1; 6555th Aerospace Test Group History, 1 July - 31 December 1975, STS Division Historical Section, pp. 2, 5, 6; 6555th Aerospace Test Group History, 1 January - 30 June 1976, STS Division Historical Section, "Division Activities;" 6555th Aerospace Test Group History, 1 July - 31 December 1976, STS Division Historical Section, "Mission."

79. 6555th Aerospace Test Group History, 1 January - 30 June 1977, STS Division Historical Section, pp. 2, 5; 6555th Aerospace Test Group History, 1 July - 31 December 1977, STS Division Historical Section, pp. 1, 6; 6555th Aerospace Test Group History, January - September 1978, pp. 9, 10; 6555th Aerospace Test Group History, October 1978 - September 1979, p. 23.

80. 6555th Aerospace Test Group History, October 1978 - September 1979, Volume I, pp. 5, 25, 26; ESMC History, 1 October 1979 - 30 September 1981, Volume I, p. 23; ESMC History, 1 October 1981 - 30 September 1982, Volume I, pp. 10, 11; ESMC History, 1 October 1982 - 30 September 1984, Volume I, pp. 3, 4.

81. ESMC History, 1 October 1981 - 30 September 1982, Volume I, p. 184; 45 SPW History, 1 January - 31 December 1992, Volume I, pp. 188, 190.

82. 45 SPW History, 1 January - 31 December 1992, Volume I, pp. 190, 191; Interview, Mark C. Cleary with Lt. Colonel Bob Thunker, Director of Operations, 45 OPG/LV, 8 September 1993.

83. 45 SPW History, 1 January - 31 December 1992, Volume I, pp. 186, 187.

84. 45 SPW History, 1 January - 31 December 1992, Volume I, p. 187; Interview, Mark C. Cleary with Captain Otis J. Campbell, 45 RANS/DOS, 9 September 1993.

85. 45 SPW History, 1 January - 31 December 1992, Volume I, pp. 187, 188.

86. ESMC History, 1 October 1981 - 30 September 1982, Volume I, pp. 186, 187, 213, 214; NASA, "Astronaut Fact Book," February 1992.

87. ESMC History, 1 October 1982 - 30 September 1984, Volume I, pp. 197, 198, 199, 200; ESMC History, 1 October 1984 - 30 September 1986, Volume I, p. 238; ESMC History, 1 October 1989 - 30 September 1990, Volume I, p. 286; NASA, "Astronaut Fact Book," February 1992.

88. ESMC History, 1 October 1984 - 30 September 1986, Volume I, pp. 237, 238, 240, 241; Crespino, "Launches," p. 46; NASA, "Astronaut Fact Book," February 1992.

89. ESMC History, 1 October 1984 - 30 September 1986, Volume I, pp. 241, 242, 243; Crespino, "Launches," p. 46.

90. ESMC History, 1 October 1984 - 30 September 1986, Volume I pp. 244, 246, 247; Crespino, "Launches," p. 46; NASA, "Astronaut Fact Book," February 1992.

91. ESMC History, 1 October 1984 - 30 September 1986, Volume I, pp. 258, 259, 260, 261; Crespino, "Launches," p. 46; NASA, "Astronaut Fact Book," February 1992.

92. ESMC History, 1 October 1984 - 30 September 1986, Volume I, pp. 261, 262.

93. ESMC History, 1 October 1988 - 30 September 1989, Volume I, pp. 334, 335, 336.

94. ESMC History, 1 October 1988 - 30 September 1989, Volume I, pp. 342, 343.

95. ESMC History, 1 October 1989 - 30 September 1990, Volume I, p. 285; NASA, "Astronaut Fact Book," February 1992.

96. ESMC History, 1 October 1989 - 30 September 1990, Volume I, pp. 285, 286, 287; NASA, "Astronaut Fact Book," February 1992.

97. ESMC History, 1 October 1989 - 30 September 1990, Volume I, pp. 287, 288; Crespino, "Launches," p. 47; NASA, "Astronaut Fact Book," February 1992.

98. ESMC/ 45 SPW History, 1 October 1990 - 31 December 1991, Volume I, pp. 298, 299, 300.

99. ESMC/45 SPW History, 1 October 1990 - 31 December 1991, Volume I, p. 305.

100. ESMC History, 1 October 1982 - 30 September 1984, Volume I, pp. 184, 185; Briefing, CSR, "Range Operation Briefing, Space Shuttle Mission STS-39," March 1991, pp. 3, 4, 5, 6, 7; ESMC History, 1 October 1990 - 31 December 1991, Volume I, pp. 305, 306.

101. ESMC/45 SPW History, 1 October 1990 - 31 December 1991, Volume I, p. 307.

102. ESMC/45 SPW History, 1 October 1990 - 31 December 1991, Volume I, pp. 313, 314.

103. Cleary, *The 6555th*, p. 159.

The Cape, Chapter 3, Section 1

Medium and Light Military Space Operations

Medium Launch Vehicle and Payload Operations

the beginning of 1971, the Cape's medium military space operations were carried out by contractors and coordinated through two of the 6555th Aerospace Test Group's divisions: the ATLAS Systems Division and the TITAN III Systems Division. Under the direction of Lt. Colonel Bobby J. Hilbert, the ATLAS Systems Division supervised ATLAS/AGENA operations on Complex 13. The Division had thirteen officers, nine airmen and five civilians assigned in January 1971, and their tasks included payload and launch vehicle test planning, integration and control. Five different contractors provided the major components and services for the ATLAS/AGENA launch system. Rocketdyne provided the ATLAS vehicle's liquid rocket engines, and General Dynamics Convair (GDC) provided the ATLAS launch vehicle and launch services. Lockheed Missile and Space Company (LMSC) provided the AGENA upper stage and assisted with its integration atop the ATLAS vehicle. General Electric operated the ATLAS MOD III ground controlled radio guidance system, and Burroughs operated the MOD III's computer system. Though General Dynamics Convair and Lockheed Missiles and Space Company had the lion's share of contractor personnel working ATLAS/AGENA operations, General Electric, Burroughs and Rocketdyne had about three dozen employees working on three different contracts for the ATLAS/AGENA program at the Cape. General Dynamics operated Complex 13 and hangars J, K, Little J and Little K. Lockheed worked at the Satellite Support Building and hangars E and AA. General Electric and Burroughs operated the MOD III Guidance Facility at the Cape. Elsewhere on the Cape, the TITAN III Systems Division's payloads branch coordinated British SKYNET II and NATO communications satellite missions with NASA and its contractors. (The NATO and SKYNET II satellites were launched on DELTA boosters under NASA's direction at Complex 17.) The Payloads Branch also handled much of the preliminary coordination with NASA on ATLAS/CENTAUR Fleet Satellite Communications (FLTSATCOM) missions before technical surveillance of the effort was transferred to the ATLAS Systems Division in June 1973. That change was relatively minor, but the 6555th cut across heavy (TITAN) and medium (ATLAS and DELTA) launch vehicle/payload lines when it complied with the 6595th Commander's request to reorganize those resources under the Satellite Systems Division and the Space Launch Vehicle Systems Division on 1 November 1975. Thereafter, the focus for medium military space operations revolved around the ATLAS/AGENA Launch Operations Branch and various project officers in the Satellite Systems Division.[1]

On 2 February 1971, A DELTA launch vehicle carrying the NATO IIB communications satellite lifted off Pad 17A on a successful mission at 2042:00 Eastern Standard Time. The flight was a NASA-directed operation, but members of the TITAN III Systems Division's payloads branch were tasked to help contractors check out the satellite and verify the NATO IIB's readiness for launch. Later in the year, the ATLAS Systems Division worked with contractors to prepare a vehicle and payload for a classified ATLAS/AGENA launch from Complex 13. Unfortunately, that mission failed after the launch vehicle malfunctioned during its flight on the evening of 4 December 1971. The next two ATLAS/AGENA missions were completed successfully on 20 December 1972 and 6 March 1973. Both missions involved classified experimental satellites which were boosted into earth orbit. In the meantime, the TITAN III Systems Division Payloads Branch coordinated installation of the Remote Vehicle Checkout Facility (RVCF), which would be used to verify a satellite's ability to receive and respond to telemetry and command signals sent from remote tracking stations in the Air Force Satellite Control Facility (AFSCF) network. Following equipment installation and system validation, the RVCF was declared "operational" on 1 September 1972.[2]

Figure 96: NATO IIB launch 2 February 1971

The Payloads Branch became the TITAN IIIC Satellite Systems Office in early 1973, and it continued to support the British SKYNET II military communications satellite program as well as half a dozen other space-related programs. Regarding the SKYNET II program, the British wanted their satellites placed in synchronous equatorial orbits at an altitude of approximately 22,300 miles. Two earlier SKYNET satellites had been launched, but only one had been orbited successfully. The United Kingdom's Ministry of Defence delayed the next SKYNET launch (SKYNET II-A) for some time, and the spacecraft did not arrive at the Cape until 11 December 1973. Payload processing in the Satellite Assembly Building went fairly smoothly, and the spacecraft was mated at the launch pad to an extended long tank DELTA launch vehicle equipped with three Castor II solid rocket motors. The SKYNET II-A payload was launched from Pad 17B on 18 January 1974 at 2039:00 Eastern Standard Time. Though the initial phase of the launch went well, the DELTA's second stage guidance system malfunctioned, and the payload was lost in space until the Air Force Satellite Control Facility discovered it in an unusable low-Earth orbit six days later. Despite efforts to save the spacecraft, the SKYNET II-A was destroyed as it reentered the atmosphere over the southwest Pacific Ocean on 27 January 1974.[3]

The next payload in the SKYNET series was SKYNET II-B, and it arrived at the Cape in 1974. Due to modifications underway at the Satellite Assembly Building, the spacecraft was diverted to NASA Hangar AE for processing. Following a successful checkout, the SKYNET II-B was transported to Pad 17B and mated to an extended long tank DELTA equipped with three Castor II solid rocket motors. The mission was launched on 22 November 1974 at 1928:00 Eastern Standard Time, and it was highly successful. Like the SKYNET A launched five years earlier, the SKYNET II-B was placed in a near synchronous equatorial orbit over the Indian Ocean. The SKYNET II-B continued to operate successfully for the next ten years.[4]

Figure 97: SKYNET B launch 19 August 1970

As we noted earlier, the ATLAS Systems Division was given responsibility for coordinating FLTSATCOM missions with NASA in June 1973. In addition to planning the first flights, the Division was expected to monitor launch vehicle and payload flight readiness despite NASA's overall direction of the launches at Complex 36. The first ATLAS/CENTAUR FLTSATCOM flight was projected to occur in 1975, but the date continued to slip through the end of 1975 until it was not expected before 1978. As a result of this circumstance, other medium launch vehicle missions continued to dominate the ATLAS Systems Division's attention (and that of its successors in other organizations) through 1978. Much of the activity centered on operations at Complex 13 and hangars E and J . Launches were few and far between, but an ATLAS/AGENA vehicle was launched from Complex 13 at 0500:00 Eastern Daylight Time on 18 June 1975. The classified payload aboard the vehicle included various scientific experiments, and it was placed in orbit successfully. The ATLAS/AGENA launch program remained dormant for nearly two years before its next classified mission was launched from Complex 13 on 23 May 1977. That mission got underway with a lift-off at 1313:00 Eastern Daylight Time. No significant problems were noted, and "user requirements were met." Another classified payload was boosted into space from Complex 13 on the evening of 11 December 1977, and the last ATLAS/AGENA mission was launched at 1945:00 Eastern Standard Time on 6 April 1978. Both missions were successful.[5]

Figure 98: SKYNET II-A launch 18 January 1974

Under the direction of Major Jerry H. Freer, the Satellite Systems Division pursued other medium military space missions with NASA. Toward the end of 1975, the Division completed final preparations for the NATO IIIA communications satellite, which was expected to arrive at the Cape on 14 January 1976. The satellite finally arrived on 15 March 1976, and it was processed in NASA's Hangar AO before it was moved to Area 60A on April 7th for Apogee Kick Motor (AKM) installation and fueling. After those actions were completed, the spacecraft was weighed. The NATO IIIA was transferred to Complex 17, where it was mated to a DELTA booster on 15 April 1976. The payload fairing was installed on April 20th, and the mission was launched successfully from Pad 17B at 1546:00 Eastern Standard Time on 22 April 1976. After the spacecraft was placed in a nominal transfer orbit, the Air Force Satellite Control Facility fired the AKM to place the NATO IIIA in a 22,300-nautical-mile near synchronous orbit at 18 degrees west longitude. All but one spacecraft system performed satisfactorily, and the mission was a success.[6]

Under the direction of Lt. Colonel Russell E. Vreeland, Jr., the Satellite Systems Division supported prelaunch processing operations for the NATO IIIB spacecraft. The NATO IIIB arrived at the Cape's Skid Strip on 15 December 1976, and it was taken to NASA's Hangar AO for checkout. The satellite's Remote Vehicle Checkout Facility testing was completed on December 17th, and its communications and electrical checks were completed on 10 January 1977. The spacecraft was transported to Area 60A on the 10th, and the AKM was installed and fueled. After the NATO IIIB was mated to the DELTA third stage, it was transported from Area 60A to Pad 17B on 20 January 1977. Following final mating and launch readiness tests at the pad, DELTA 128 lifted off at 1950:00 Eastern Standard Time on 27 January 1977. The payload was placed in a highly elliptical synchronous transfer orbit. The Air Force Satellite Control Facility at Sunnyvale, California fired the spacecraft's AKM two days later to circularize the NATO IIIB's orbit. On-orbit checkout proceeded smoothly, and the satellite took up its station at 135 degrees west longitude, over the eastern Pacific Ocean.[7]

Figure 99: NATO IIIA launch 22 April 1976

After years of preparation and scheduling delays, the first FLTSATCOM spacecraft finally arrived at the Skid Strip on 2 December 1977. The FLTSATCOM F-1 was transported to the Satellite Assembly Building (SAB) for its receiving inspection, and pressure/leak tests were completed successfully a few days later. The spacecraft's Reaction Control System (RCS) was tested from 7 through 13 December 1977, and the Apogee Kick Motor's safe and arm devices were checked out by December 21st. A special 200-hour-long "burn-in" test was required to check for a transistor problem discovered on the FLTSATCOM F-2 spacecraft (still at the factory), but that test was completed on December 22nd with no problems. Following the last of its receiving inspections, the spacecraft was transported to the Satellite Assembly and Encapsulation Facility on 5 January 1978. Processing was delayed about two weeks due to another F-2 spacecraft anomaly (e.g., bad resistors and transistors), but processing continued after the SAMSO Commander authorized use of the existing components on the F-1. The spacecraft was moved to Pad 36A on 28 January 1978, and it was mated to the ATLAS/CENTAUR (AC-44) launch vehicle. Following its final compatibility and launch readiness tests, the vehicle lifted off at 1617:00 Eastern Standard Time on 9 February 1978. The spacecraft entered the proper transfer orbit approximately 1510 seconds after launch. The Air Force Satellite Control Facility fired the spacecraft's AKM on the fifth orbit (i.e., about two days later), and the subsequent "burn" placed the FLTSATCOM F-1 in a nearly geostationary orbit 22,300 nautical miles above the equator at 100 degrees west longitude. From that location, the F-1 provided 23 channels of super high frequency communications to meet vital needs in the Defense Department and the Presidential Command Network. The mission was a success.[8]

The NATO IIIC spacecraft was the next medium military payload to be processed at the Cape, and it arrived at the Skid Strip on 18 September 1978. The satellite's validation processing began fairly quickly in the fall of 1978, but it hit a snag when two signal strength anomalies were discovered in the NATO IIIC's downlink telemetry system. Extensive troubleshooting traced the problem to a radio frequency switch, which was removed and returned to Ford Aerospace and Communications Corporation for rework and retesting. A replacement switch and a new power amplifier module were installed in the spacecraft, and processing continued. Following mating with its DELTA third stage, the NATO IIIC was taken out to Pad 17B and mated to its launch vehicle. The mission was launched at 1946:00 Eastern Standard Time on 18 November 1978, and the spacecraft entered the prescribed 100 x 19,232-nautical-mile transfer orbit successfully. Two days later, the Air Force Satellite Control Facility fired the spacecraft's AKM to circularize the NATO IIIC's orbit. On-orbit testing was routine, and the satellite was released to NATO in February 1979. The NATO IIIC was placed in orbital storage initially, where it served as a backup satellite for the other two NATO III satellites in the constellation.[9]

As we noted earlier, the FLTSATCOM F-2 had more than its share of transistor and resistor problems at the factory in 1977 and 1978, but the spacecraft finally arrived at the Cape on 30 March 1979. The F-2's pressurization and leak tests were completed at the Satellite Assembly Building on April 4th. A systems test followed, and the spacecraft was moved to the Remote Vehicle Control Facility to verify its telemetry systems and responsiveness to commands. Following those tests, the spacecraft was transferred to the Spacecraft Assembly and Encapsulation Facility on April 17th. The F-2's Apogee Kick Motor was installed shortly thereafter. A minor pressure problem delayed the completion of final fueling and pressurization until April 20th, but final mechanical closeout and encapsulation tasks were completed on April 25th. The payload was moved out to Complex 36 and mated to its ATLAS/ CENTAUR vehicle. Following an on-stand function and certification test on April 27th and a practice countdown on April 28th, the vehicle was readied for launch. Though the countdown on May 3rd had to be aborted at T minus 90 minutes due to inoperative thermistors on the Apogee Kick Motor, the integrity of the AKM's igniter circuitry was tested and reconfirmed, and the countdown was recycled the following day. The lift-off was recorded at 1557:00 Eastern Daylight Time on 4 May 1979. The spacecraft was placed in a proper transfer orbit and (later) geosynchronous orbit. On-orbit testing was completed successfully in mid-July 1979.[10]

Figure 100: Prelaunch Test for the FLTSATCOM F-1 mission
9 February 1978

Figure 101: FLTSATCOM F-2 launch
4 May 1979

The third FLTSATCOM spacecraft (FLTSATCOM F-3) was scheduled for launch on 4 December 1979, but the mission was delayed about six weeks to correct problems in the launch vehicle's hydraulic and pneumatic systems. In the meantime, the payload could not be processed through the Satellite Assembly Building due to scheduling conflicts, and it had to be diverted to NASA's Hangar AM instead. Those workarounds aside, the F-3 was launched from Pad 36A on 17 January 1980 at 2026:00 Eastern Standard Time. Two days later, the Air Force Satellite Control Facility fired the spacecraft's Apogee Kick Motor to place the FLTSATCOM spacecraft in its final equatorial orbit 22,250 miles above the Atlantic Ocean. The mission was successful.[11]

Two more FLTSATCOM satellites were launched on ATLAS/CENTAUR vehicles from Pad 36A in October 1980 and August 1981. The FLTSATCOM-D was launched on 30 October 1980 at 10:54 p.m. (local time), and it was placed in near geosynchronous orbit over the Pacific Ocean. The FLTSATCOM-E spacecraft was launched on 6 August 1981 at 0316:00 Eastern Daylight Time. Though the FLTSATCOM-E's orbit was eccentric and lower than desired, ground controllers eventually coaxed the satellite into a satisfactory orbit. Unfortunately, the satellite had suffered considerable damage to its solar panels during the flight into space, and its UHF receive antenna could not be deployed. The FLTSATCOM-E remained in orbit as a spare satellite for the FLTSATCOM network.[12]

Figure 102: FLTSATCOM F-3 launch
17 January 1980

Under the direction of Lt. Colonel James E. Stangel, the Satellite Systems Division prepared in the early 1980s to support a wide variety of military spacecraft programs including: 1) the DSCS III communications satellite program, 2) the DSP satellite program, 3) follow-on processing operations for the SKYNET and NATO III satellite programs and 4) the introduction of NAVSTAR Global Positioning Satellite (GPS) processing operations at the Cape. (As the reader will recall, DSCS and DSP missions were discussed in connection with heavy military space operations in Chapter II, so only the other programs deserve further comment.) Concerning the NATO III program, the Spacecraft Program Office announced in November 1982 that the NATO IIIE spacecraft would not be procured, but the NATO IIID would be launched as a "gap-filler" until enhanced NATO IV communications satellites entered service sometime after 1987. The Satellite Systems Division completed its review of SKYNET IV documentation in 1983, and the Memorandum of Agreement and Program Requirements Document were signed by the British, the Air Force and NASA later in the year. By the end of 1983, the first SKYNET IV payload (SKYNET IV-A) was scheduled to be launched on the Space Shuttle Atlantis in December 1985, and the SKYNET IV-B was on Columbia's manifest for a June 1986 Shuttle launch. The Division's GPS program efforts in 1983 and 1984 focused on military construction projects to modify Cape facilities to support the new satellite program. Though GPS satellites were supposed to be launched aboard the Space Shuttle, they were diverted to DELTA II boosters after the Challenger disaster in January 1986. Let's look at each of these developments in turn-the NATO IIID mission, SKYNET IV developments and the evolution of the NAVSTAR GPS program at the Cape in the 1980s and early 1990s.[13]

Figure 103: NATO IIID launch 14 November 1984

The NATO IIID mission suffered a number of setbacks and delays in 1984, but it was launched after several reschedulings late in the year. The spacecraft's primary and backup apogee kick motors were inspected at the Cape's ordnance area in October 1983. As plans to launch the satellite in May 1984 continued, Space Division rightly suspected that the mission would be delayed by the late arrival of essential parts from spacecraft subcontractors. The spacecraft eventually arrived at the Cape on 22 August 1984, and the first two stages of its DELTA launch vehicle were erected on Pad 17A in anticipation of an 18 October launch. Unfortunately, launch base processing had to be halted after a Travelling Wave Tube Amplifier (TWTA) problem surfaced at the factory in October to cast doubt on the reliability of TWTA assemblies everywhere. The spacecraft's TWTAs were removed from its communications system and tested. After passing the tests, the TWTAs were reinstalled, and the spacecraft was mated to its DELTA third stage on November 4th. The NATO IIID was transferred from the DELTA Spin Test Facility (DSTF) to the launch pad, and the spacecraft was mated to the DELTA launch vehicle shortly thereafter. The final spacecraft closeout and fairing installation operations were completed in time for a launch attempt on November 12th, but the countdown was scrubbed on that date due to high wind shear in the upper atmosphere. The countdown was picked up again at 1849Z (Greenwich Mean Time) on November 13th, and it proceeded uneventfully to lift-off at 0034Z on 14 November 1984. The spacecraft was placed in the desired 225 x 19,323-nautical-mile transfer orbit, and the Air Force Satellite Control Facility fired the NATO IIID's Apogee Kick Motor to circularize the spacecraft's orbit two days later. On-orbit testing was completed successfully, and the NATO IIID entered service as a spare for the NATO III satellite constellation.[14]

Figure 104: Commercial TITAN III launch of SKYNET IV spacecraft
1 January 1990

Though the SKYNET IV-A and SKYNET IV-B were supposed to be launched via Space Shuttle in December 1985 and June 1986, the two missions were slipped in 1985 to June and December 1986 before the Challenger disaster derailed them completely. Eventually, the SKYNET spacecraft were diverted to other launch vehicles. On 11 December 1988, the SKYNET IV-B was launched on an ARIANE 4 launch vehicle from the European Space Agency's ELA-2 launch pad in Kourou, French Guiana. Lift-off occurred at 0033:41Z, and the spacecraft was injected into the proper orbit. Another SKYNET IV spacecraft was launched along with a Japanese communications satellite from Complex 40 on the first Commercial TITAN III vehicle on 1 January 1990. That mission had the dubious distinction of being one of the most "scrubbed" missions in the Cape's history. Following the first launch attempt, which was scrubbed on 7 December 1989 due to an error in guidance software, countdowns were started on seven subsequent occasions between December 8th and the end of the month. Finally, on 31 December 1989, the countdown was picked up at 1607Z, and it proceeded uneventfully to a successful launch at 0007:01Z on January 1st. A third SKYNET IV communications satellite was launched successfully from Kourou by the European Space Agency at 2246:04Z on 30 August 1990.[15]

Medium and Light Military Space Operations

Evolution of the NAVSTAR Global Positioning System and Development of the DELTA II

Communications satellites continued to play an important role in military space operations, but the NAVSTAR GPS program opened up a whole new field for space support operations at the Cape in the 1980s: the launching of satellites to provide highly accurate three-dimensional ground, sea and air navigation. The U.S. Navy and Air Force began the effort in the early 1960s with a series of studies and experiments dealing with the feasibility of using satellite-generated radio signals to improve the effectiveness of military navigation. After ten years of extensive research, the services concluded that Defense Department requirements would be best served by a single, highly precise, satellite-based Global Positioning System (GPS). In December 1973, the Defense Navigation Satellite System (later know as NAVSTAR GPS) entered its concept validation phase. The technology necessary to field the GPS was confirmed during that phase, and four advanced development model Block I NAVSTAR satellites were launched on ATLAS F boosters from Vandenberg's Space Launch Complex 3 (East) between 22 February and 11 December 1978. Two more Block I satellites (NAVSTAR 5 and NAVSTAR 6) were launched on ATLAS F boosters from Complex 3 (East) on 9 February and 26 April 1980. By the end of 1980, the NAVSTAR GPS constellation was arranged in two orbital planes of three satellites each, orbiting Earth at an altitude of approximately 10,900 nautical miles. Following the GPS development phase in the early 1980s, the Air Force planned to procure and deploy a constellation of 24 Block II GPS satellites via Space Shuttle by the end of 1987. Funding cuts in 1980 and 1981 reduced the planned constellation to 18 Block II satellites and added a year to their deployment, but the program continued to move ahead.[16]

As preparations for the Block II GPS satellite program continued, a Block I replenishment satellite was launched on an ATLAS E booster from Complex 3 (East) on 18 December 1981. Unfortunately, a hot gas generator on one of the ATLAS booster's main engines failed shortly after lift-off, and the vehicle crashed about 150 yards from the pad. The next replenishment satellite launch was postponed while ATLAS engines were refurbished and test-fired in 1982, but the mission was finally launched successfully from Complex 3 (West) on 14 July 1983. The satellite (NAVSTAR 8) replaced NAVSTAR I in the 240-degree orbital plane of the GPS constellation. The last three Block I satellite missions (NAVSTARs 9, 10 and 11) were launched on ATLAS E boosters from Complex 3 (West) on 13 June 1984, 8 September 1984 and 8 October 1985. All three launches were successful, and the satellites performed as planned.[17]

On 20 May 1983, the Air Force signed a $1.2 billion five-year fixed price contract with Rockwell International's Satellite Systems Division for the procurement of 28 Block II NAVSTAR GPS satellites. The contract was the Defense Department's first multi-year procurement of production model satellites, and, as a "block buy," it reduced the average cost of a Block II satellite by about 24 percent (e.g., $40,600,000 versus $52,000,000). The ambitious production schedule called for the first spacecraft in Fiscal Year (FY) 1984. Six more Block II satellites were expected to be built in FY 1985. Nine more satellites were anticipated in FY 1986, followed by eight more GPS spacecraft in FY 1987. The last four Block II satellites were scheduled for assembly in FY 1988. In the meantime, the Air Force and McDonnell Douglas reached agreement in the summer of 1983 on a separate $169,400,000 contract to provide payload assist modules (e.g., the PAM-DII) to boost 28 Block II satellites from the Shuttle's orbit into their own elliptical transfer orbits. NAVSTAR Support facilities were the next order of business: as Block II satellites and PAM-DII modules arrived at the Cape, they were to be mated to each other in a NAVSTAR Processing Facility (NPF) modified expressly for that purpose. One of the Cape's old MINUTEMAN missile assembly buildings-MAB-2-was chosen as the site for the facility, and the NPF design was completed in April 1984. In June 1984, a $3,800,000 construction contract was awarded to W & J Construction Corporation of Cocoa, Florida to build the facility. Modifications to a Propellant Servicing Facility (PSF) and a Motor Inert Storage (MIS) building raised the overall cost of the project to approximately $4,500,000. The ground breaking ceremony for the NAVSTAR Processing Facility was held on 5 July 1984, and construction was completed in July 1985.[18]

Testing of the first Block II satellite was well underway in 1985, but the NAVSTAR II satellite program was already markedly behind schedule. Serious workmanship problems began to surface in the electronic flight boxes supplied by Rockwell's Strategic Defense and Electro-Optical Systems Division in Anaheim, California, and this situation threatened more delay. A "tiger team" was organized to reopen, inspect and repair all the flight boxes earmarked for the program, but, despite the company's best efforts, the first production Block II would not be ready for launch by the date indicated in the contract (e.g., 22 August 1986). By the fall of 1985, the first Block II mission had to be rescheduled from October 1986 to January 1987. On a brighter note, the first PAM-DII payload assist module and the Block II NAVSTAR qualification satellite (GPS-12) arrived at the Cape on 22 January and 21 March 1986 respectively. Under the NAVSTAR II Pathfinder program, the PAD-DII and the qualification satellite were processed to verify the readiness of facilities and ground equipment to support NAVSTAR II operations. The PAM-DII was built up and checked out by April 1st, and a Combined Systems Test (CST) was completed on the satellite at the NAVSTAR Processing Facility on 10 April 1986. The satellite and PAM-DII were mated and checked out successfully on 16 May 1986.[19]

Figure 105: NAVSTAR II Payload and PAM-D Upper Stage

Following the Challenger disaster in January 1986, the GPS Program Office replanned the first eight Block II satellites for flights on the new Medium Launch Vehicle (the DELTA II) in lieu of the Space Shuttle. The Shuttle was still being considered for the next eight GPS missions (e.g., two spacecraft per Shuttle flight), but its prospects for NAVSTAR missions dimmed as pressure increased to replenish the GPS constellation quickly and meet user needs. In August 1987, the Secretary of the Air Force decided to transfer all but four of the 28 NAVSTAR II satellites from the Shuttle to the DELTA II. Only two NAVSTAR II satellites remained on the Shuttle's manifest by the end of FY 1988, and those last two spacecraft were reassigned to the DELTA II in the spring of 1989. In any event, the first NAVSTAR II launch slipped two years to January 1989 (though some optimists thought it might occur as early as October 1988). Hopefully, 21 GPS satellites would be in orbit by January 1991. In the meantime, arrangements were made to store Block II satellites and suspend work on batteries, apogee kick motors and other components with short shelf lives. Those necessary actions added $153 million to the Block II contract.[20]

The changes in launch vehicles and schedules for the NAVSTAR II launch program affected the Air Force's contract with McDonnell Douglas for 28 PAM-DII upper stages. The PAM-DII was designed to boost a NAVSTAR II satellite into transfer orbit after it had been released from the Shuttle at an altitude of approximately 160 nautical miles. The DELTA II would use a different upper stage-a newly configured PAM-D-to boost the NAVSTAR II payload into transfer orbit. Following the decision to go with the DELTA II, the Air Force cancelled the PAM-DII contract. The U.S. Government only accepted 16 PAM-DIIs it had already purchased.[21]

As noted in Chapter I, Space Division awarded the Medium Launch Vehicle (MLV) contract to McDonnell Douglas Astronautics Company on 21 January 1987. Together with its options, that DELTA II contract was valued at $669,000,000. McDonnell Douglas also had at least four firm orders from non-military customers to launch DELTA II vehicles on commercial missions. However, unlike earlier commercial arrangements, the company would no longer be under contract to NASA. Under the new Commercial Expendable Launch Vehicle program encouraged by President Reagan since 1983, McDonnell Douglas would be responsible for producing, marketing and launching its commercial DELTA IIs. The Air Force would be responsible for ensuring safety and environmental standards for commercial as well as military launches, but McDonnell Douglas would have greater responsibility in meeting those standards (including quality control). Both launch pads (17A and 17B) would be equipped to handle commercial and Defense Department missions. McDonnell Douglas and its subcontractors were soon hard at work preparing the pads for the new DELTA II vehicles.[22]

Figure 106: Mobile Service Towers at Complex 17 (looking north)
October 1988

The DELTA IIs would be taller than earlier DELTA vehicles (e.g., 130 feet versus 112 feet), and one of the contractor's first tasks was to raise Complex 17's Mobile Service Tower (MST) 20 feet to accommodate the DELTA II's longer stages. Other modifications revolved around Pad 17A initially because Pad 17B was committed to the DELTA 181 mission which was scheduled to be launched in February 1988. While DELTA 181's first stage was being erected in October 1987, the steelwork subcontractor (Butler Construction Company) began falling behind on Pad 17A. Seven weeks later, Butler was seven weeks behind schedule. Though bureaucratic delays and changes in McDonnell Douglas' job orders were cited for the holdup, the problem seemed to gravitate toward steel fabrication delays at MET-CON's offsite shop. Whatever the real cause of the delay, MET-CON was willing to make up the lost time "if paid to do so." A recovery schedule was created to get Pad 17A's modifications back on track and completed by mid-February 1988. In the meantime, offsite steelwork was accelerated to avoid further delays once work on Pad 17B began. The contractor expanded his workweek to ten hours per day, seven days a week in January 1988. Pad 17's modifications were essentially complete by mid-April 1988, and Pad 17B's work was on schedule. The contractor's remarkable recovery was due in large part to having most of the offsite prefabrication work completed before modifications on Pad 17B began.[23]

Unfortunately, trouble loomed from a different quarter in July 1988: McDonnell Douglas ran into trouble getting some fiber optic equipment it ordered for Pad 17A, and the first DELTA II launch was rescheduled from 13 October 1988 to 8 December 1988. There was still plenty of work to be done, and the contractor's people extended their workdays from 6:00 a.m. to 9:00 p.m. to complete the tasks that remained. (In some instances, people returned as early as 4:00 a.m. the next morning to meet their schedules from the previous day.) Test discrepancies in McDonnell Douglas' plant also delayed the first launch somewhat longer, but the first DELTA II's first stage was erected on Pad 17A by 2 November 1988. The vehicle's interstage was installed at the pad on November 5th, and the solid rocket motors were mated to the vehicle a few days later. The range contractor and ESMC's engineers completed launch data connections between the blockhouse and Pad 17 around the middle of November 1988. On 24 January 1989, command and telemetry verification tests confirmed reliable links between Sunnyvale and Colorado Springs for the upcoming NAVSTAR II GPS mission. Following final prelaunch tests, the countdown was picked up on 12 February 1989, but it was scrubbed at 1827Z due to excessive high altitude winds. The countdown was picked up again on February 14th, and lift-off was recorded at 1829:59.988Z on 14 February 1989. The first DELTA II placed the first NAVSTAR II GPS payload into the proper transfer orbit. The mission was a success.[24]

Medium and Light Military Space Operations

DELTA II Processing and Flight Features

Before we look at the remaining NAVSTAR II GPS missions, a short review of DELTA II processing and flight highlights is in order. Like the TITAN and ATLAS lines of vehicles, the DELTA line was built on major components supplied by several different contractors. McDonnell Douglas built the basic core vehicle and supplied fairing materials at its plant in Huntington Beach, California, but it shipped them to another plant in Pueblo, Colorado for further assembly and/or match ups with other contractors' components. Rocketdyne provided the DELTA's main engine, and Aerojet supplied the vehicle's second stage engine. DELCO supplied the inertial guidance system, and Morton Thiokol built the strap-on solid rocket motors used for the basic Model 6925 DELTA II vehicle. (Hercules built the Graphite Epoxy Motors [GEMs] used on the more powerful Model 7925.) Once the core vehicle stages and fairing were assembled, they were transported by truck to the Cape and received at Hangar M. After their initial inspection at Hangar M, the first and second stages were taken to the DELTA Mission Checkout Operations Area, where they received telemetry and controls checks, flight simulations and dual composite testing. The first and second stages were transported to the Horizontal Processing Facility (HPF) in Area 55 for destruct system installation. Following processing at the HPF, both stages were moved to Complex 17 and erected.[25]

Figure 107: First DELTA II erection (left) and SRM mate (right) at Pad 17A November 1988

Figure 108: Solid Rocket Motor Operations in Area 57

The strap-on solids and the new DELTA-configured PAM-D followed different processing flows. The DELTA II's Morton Thiokol or Hercules solid rocket motors were trucked to the Cape's Area 57 initially. Once there, the solids were: 1) inspected, 2) checked for leaks and flaws in the solid propellant and 3) built up with the required destruct harnesses and nose cones. Finally, the assembled solid rocket motors were placed on transporters and moved out to Complex 17. The DELTA II's PAM-D motor was received at an ordnance storage area where it was inspected, placed in a cold chamber and cold-soaked. Later, the motor was transferred to the Non-Destructive Test Laboratory (NDTL) where it was x-rayed and placed in a shipping container for transport to the NAVSTAR Processing Facility. At the NPF, the motor was assembled into a complete PAM-D by adding a payload attachment fitting equipped with a fueled nutation control system and a spin table. Following assembly, the PAM-D was spin-balanced at NASA's Explosive Safe Area 60 (on the Cape). The PAM-D was returned to the NAVSTAR Processing Facility and mated to the spacecraft. The payload was installed in a McDonnell-Douglas payload container, loaded on a transporter, and driven out to the launch pad.[26]

At Complex 17, the entire process came together to create a complete DELTA II launch vehicle. The interstage and payload fairing were brought out to the launch pad from Hangar M. The first stage, solid rocket motors, interstage and second stage were erected and mated at the pad, and the payload fairing was secured in the Mobile Service Tower. The solid rocket motors were aligned, and umbilicals were installed. Electrical and mechanical qualification checks were accomplished about a month before the launch, and subsystem checks continued as the contractor prepared to mate the vehicle with its payload about nine days before lift-off. Ordnance connections and safety checks continued during the last week on the pad, and the vehicle was prepared for launch.[27]

Figure 109: PAM-D Motor

Figure 110: Spacecraft/PAM-D Premate Closeout

The first DELTA IIs were Model 6925 vehicles equipped with nine Morton Thiokol Castor IVA solid rocket motors, but a more powerful version-the Model 7925-made its debut in November 1990. The Model 7925 was equipped with nine Hercules Graphite Epoxy Motors (GEMs) and a more powerful Rocketdyne main engine to boost heavier Block IIA NAVSTAR satellites into orbit. Despite the differences in payload and power, all DELTA II/NAVSTAR II missions had several flight sequences in common. Following lift-off, the DELTA II rolled over into a flight azimuth of 112 degrees for the Model 6925 flights and 110 degrees for the Model 7925 missions. Only six of the DELTA II's strap-on solid rocket motors or GEMs fired at lift-off, but the last three motors fired about one minute after launch (e.g., 59 seconds after launch for the Castor IVAs and 65 seconds after launch for the GEMs). The first six solids separated from the vehicle two seconds later, and the last three motors continued to burn for about one minute before they were jettisoned. The first stage cutoff occurred between 4 minutes and 20 seconds and 4 minutes and 24 seconds after launch (i.e., the Model 6925 burned four seconds longer), and the first and second stages separated eight seconds later. The second stage's Aerojet ITIP engine ignited six seconds after that, and the payload fairing was jettisoned between 11 and 14 seconds later. Following the vehicle's entry into a low-Earth parking orbit, the second stage's engine ceased firing, and the vehicle and its payload went through a "slow roll" to distribute temperature caused by solar radiation. After the vehicle reoriented itself, it coasted until either: 1) the second stage's engine fired once again to assist the PAM-D (third stage) in placing the payload into the prescribed transfer orbit, or 2) the PAM-D ignited to place the payload in the prescribed transfer orbit. In either case, the spacecraft separated from the vehicle about 25 minutes into the flight, and the Apogee Kick Motor was used along with the satellite's Reaction Control System (RCS) to circularize the NAVSTAR's orbit in one of at least six orbital planes approximately 10,900 nautical miles above Earth.[28]

Medium and Light Military Space Operations

NAVSTAR II Global Positioning System Missions

Following the first NAVSTAR II GPS mission on 14 February 1989, 16 highly successful NAVSTAR II missions were launched from Complex 17 between 10 June 1989 and the end of 1992. The first eight of those missions involved Model 6925 launch vehicles, and the remaining eight flights featured Model 7925s equipped with GEMs. Though all those missions were ultimately successful, countdown operations did not always lead to launches. The NAVSTAR II-2 mission lifted off Pad 17A at 2230:01Z on 10 June 1989, but only after five earlier countdowns were attempted and scrubbed in May and June 1989. The NAVSTAR II-3 mission was scheduled for launch on 11 August 1989, but it was delayed 24 hours after Pad 17A suffered a lightning strike. The NAVSTAR II-3's first launch attempt on August 12th was scrubbed due to a weather constraint violation at launch time, and a second countdown was required before the vehicle lifted off Pad 17A at 0557:59Z on 18 August 1989. The NAVSTAR II-4 mission was launched from Pad 17A on its first launch attempt at 0931:01Z on 21 October 1989, but the NAVSTAR II-5 mission was scrubbed on 10 December 1989 due to a helium problem in the vehicle's second stage. The NAVSTAR II-5 mission lifted off Pad 17A at 1810:01Z on 11 December 1989. The NAVSTAR II-6 mission lifted off Pad 17A on its first launch attempt at 2255:01Z on 24 January 1990, but the first countdown for the NAVSTAR II-7 mission was scrubbed on 21 March 1990 due to unacceptable upper level wind conditions. The mission was rescheduled, and NAVSTAR II-7 was launched successfully from Pad 17A at 0245:01Z on 26 March 1990. The NAVSTAR II-8 payload was launched after an uneventful countdown at 0539:00Z on 2 August 1990, and a Model 6925 boosted the NAVSTAR II-9 mission into orbit from Pad 17A at 2156:00Z on 1 October 1990.[29]

Figure 111: DELTA II Payload Encapsulation at Pad 17A January 1989

The Model 7925 made its debut with a successful lift-off from Pad 17A at 2139:01Z on 26 November 1990. Apart from a seven-minute extension to evaluate wind data, the countdown was largely uneventful. The vehicle placed its NAVSTAR II-10 payload into the prescribed orbit. The NAVSTAR II-11 mission was scrubbed on 3 July 1991 due to the loss of a satellite communication link with the range tracking station on Ascension Island. The countdown was picked up again, and the GEM-equipped

DELTA II boosted its NAVSTAR II-11 payload and the LOSAT-X payload off Pad 17A at 0232:00Z on 4 July 1991. Both payloads were released successfully. The first launch attempt for the NAVSTAR II-12 mission was aborted at 2322:00Z on 18 February 1992 due to heavy cloud cover, and the second countdown was scrubbed at 2310:00Z on February 19th for weather problems. The third and final attempt for that mission was overshadowed by bad weather and punctuated with a brief computer synchronization problem, but the final countdown on February 23rd was otherwise uneventful. The weather improved, and the DELTA II lifted off Pad 17B successfully at 2229:00Z on 23 February 1992.[30]

Figure 112: DELTA II Model 7925 launch 26 November 1990

The next NAVSTAR II mission also required three separate countdowns before its DELTA II Model 7925 vehicle lifted off Pad 17B on 10 April 1992. The NAVSTAR II-14 mission was launched from Pad 17B on the first launch attempt at 0920:01Z on 7 July 1992. Similarly, the NAVSTAR II-15 mission was launched from Pad 17A on its first launch attempt at 0857:00Z on 9 September 1992. The NAVSTAR II-16 countdown on November 7th was scrubbed at 0105Z due to a vehicle misfire, and the mission's next countdown on November 20th was scrubbed at 0005Z due to weather constraints. In contrast to the first two attempts, the countdown for the NAVSTAR II-16 launch on November 22nd went smoothly. There were no unscheduled holds, and the vehicle lifted off without incident at 2354:00Z on 22 November 1992. The final NAVSTAR II mission of 1992 was scrubbed on December 16th due to a vehicle anomaly which occurred during the final liquid oxygen filling procedure. The second countdown on December 18th was successful, and the DELTA II lifted the NAVSTAR II-17 payload off Pad 17B at 2216:00Z on that date. The flight was the seventeenth in a unbroken series of successful missions for the NAVSTAR II GPS program.[31]

The DELTA II/NAVSTAR II launch program was one the Air Force's greatest success stories at the Cape, and it is only fitting that we should note the impact of those launches on the NAVSTAR GPS constellation in the late 1980s and early 1990s. Of the ten Block I NAVSTAR spacecraft orbited before November 1985, seven were still in operation when the first NAVSTAR II mission was launched in February 1989. In 1988, the Air Force and the Joint Requirements Oversight Committee of the Joint Chiefs of Staff supported a return to the original plan of a 24-satellite GPS constellation. On 27 February 1989, the Air Force directed Air Force Systems Command to establish and maintain a GPS constellation of 21 primary satellites, plus active orbiting spares (i.e., a 24-satellite constellation). The last two NAVSTAR II satellites were removed from the Space Shuttle's payload manifest in the spring of 1989, so the entire weight of the expanded constellation fell on the shoulders of the DELTA II launch program.[32]

The GPS Program Office hoped to have five NAVSTAR II satellites in orbit by the end of September 1989, but only three of those spacecraft had been launched by that time. Since twelve Block II satellites would be needed to give the GPS constellation its first worldwide two-dimensional navigation capability, planners estimated that capability could not be achieved before the spring of 1991. In point of fact, six more NAVSTAR II satellites were launched over the next year, and Iraq's invasion of Kuwait in August 1990 provided additional incentive for McDonnell Douglas and the Air Force to rise to the challenge. NAVSTAR II-9 (the last of the six NAVSTARs mentioned) lifted off on 1 October 1990, and it was placed in orbit over the Middle East. The satellite's on-orbit testing program was completed in record time, and NAVSTAR II-9 was turned over to Air Force Space Command on 24 October 1990. NAVSTAR II-10 was launched successfully on 26 November 1990. With II-10 in operation, the GPS network provided two-dimensional coordinates with an average accuracy of 4.5 meters during DESERT STORM. The NAVSTAR system's three-dimensional accuracy averaged 8.3 meters during the war. The GPS Program Office hoped to launch five Block IIA NAVSTAR spacecraft by October 1991, but component problems associated with the new design caused lengthy delays. Only two Block IIA missions were launched by October 1991, but five more Block IIA launches were completed by the end of 1992. The constellation was well on its way to full operational status by the beginning of 1993.[33]

Medium and Light Military Space Operations

Strategic Defense Initiative Missions and the NATO IVA Mission

In addition to the NAVSTAR II missions, there were four Strategic Defense Initiative (SDI) missions launched from Complex 17 in the mid-to-late 1980s and 1990. The first of those experimental missions was DELTA 180. It involved the placement of a DELTA second stage and a Payload Adapter System (PAS) in two 120-nautical-mile-high time-synchronous orbits with slightly different inclinations. The overall objectives of the mission were to: 1) gather optical spectral data from rocket propulsion sources and 2) confirm guidance and navigation algorithms. A DELTA Model 3920 vehicle was used for the booster, and the countdown proceeded smoothly to DELTA 180's lift-off from Pad 17B at 1508:01Z on 5 September 1986. According to the post-launch debriefing report submitted by the Eastern Test Range organization after the mission, DELTA 180 was "a tremendous success both from a launch and orbital support standpoint." The mission marked the first use of an expendable launch vehicle in support of an SDI mission at the Cape.[34]

Figure 113: DELTA 180 launch 5 September 1986

Figure 114: DELTA 181 launch 8 February 1988

The next DELTA/SDI mission from the Cape was DELTA 181. Like the DELTA 180 mission, DELTA 181 involved the placement of a DELTA second stage in low-Earth orbit, but it included the deployment of two payloads as the object of a series of experiments in orbit. One of the payloads was a plume generator package, and the other was a science package containing eight test objects and four reference objects. A sensor module, consisting of a command and data handling system and seven scientific experiments, remained with the DELTA's second stage to scan elements of both packages once they were deployed. The sensor module was equipped with ultraviolet, infrared, radar and laser sensors to gather a tremendous amount of data on the "signatures" generated by the deployed payloads. That data was transmitted via two wideband telemetry downlinks to stations on the ground. Data from approximately 100 ground-based sources funneled into the Cape via communications satellites. The mission required more than 200 radar tracking maneuvers over a period of two days, and recorded data continued to come in for about ten days after the experimental portion of the mission was completed. Put simply, DELTA 181 presented the Eastern Range with one of the most complex support missions in history.[35]

Concerning the launch itself, the countdown for the DELTA 181 mission was picked up at 1557Z on 8 February 1988. It proceeded smoothly to lift-off at 2207:00Z. As the DELTA Model 3910 booster climbed away from Pad 17B, the launch vehicle rolled out of its 115-degree launch azimuth into a flight azimuth of 94 degrees. The DELTA's second stage was injected into a 90 x 120-nautical-mile orbit inclined 28.7 degrees to the equator. Data was collected during the first seven orbits, then the second stage boosted itself into a 171-nautical-mile circular orbit during its eighth revolution. During the ninth and tenth orbits, the sensor module's telemetry and tape recorder systems were verified. Over the next ten days, the sensor module transmitted its recorded data to designated ground stations in the DELTA 181 support network. The mission was highly successful.[36]

Figure 115: DELTA 183 on Pad 17B
15 March 1989

The next SDI mission was DELTA 183. It involved the DELTA STAR spacecraft, which contained a suite of sensors, a command and data handling system, seven scientific instruments and a Laser Illumination Detection System (LIDS). Like the DELTA 181 mission's data, DELTA 183 data was transmitted via telemetry downlink to stations on the ground. The primary objectives of the mission were to: 1) observe the DELTA's second stage rocket "burns" in various background environments and 2) record and transmit data about those background environments in general. The DELTA 183 launch was attempted for the first time on 15 March 1989, but the countdown was scrubbed due to launch vehicle and spacecraft problems. The second countdown was attempted on 24 March 1989, and it went well. The DELTA Model 3920 booster lifted off Pad 17B at 2150:49Z on the 24th, and the vehicle rolled into a flight azimuth of 60 degrees. During the first phase of the mission, the DELTA's second stage placed the DELTA STAR into a low-Earth orbit inclined 47.7 degrees to the equator. After coasting for approximately one-half revolution, the second stage fired again to circularize its orbit at an altitude of 269 nautical miles. The DELTA STAR spacecraft then separated from the second stage, and the second stage performed an evasive maneuver. The DELTA STAR observed the second stage's de-orbit burn and reentry into the atmosphere. During the second phase of the mission, the DELTA STAR made observations and transmitted data to ground stations over the next several months.[37]

A DELTA II Model 6925 launch vehicle was used to boost two more SDI payloads into orbit from Pad

17B on 14 February 1990. The first payload, LOSAT L, was designed to measure the absolute intensity of low energy ultraviolet, visible and infrared laser beams transmitted from a ground site to a target satellite. The second payload, LOSAT R, validated ground-based laser relay technology in the areas of beam stabilization, pointing and beacon tracking. Following lift-off at 1615:00Z on February 14th, the DELTA II rolled into a flight azimuth of 75 degrees and accomplished the "dogleg" maneuvers required to inject the payloads into their proper orbits. The LOSAT L was injected into a 546-kilometer circular orbit, and it was turned over to the U.S. Naval Research Laboratory to support various SDI studies. After the LOSAT L payload was released, the DELTA II's second stage completed a retrograde burn before releasing the LOSAT R payload into a 470-kilometer circular orbit. The U.S. Air Force Weapons Laboratory in Albuquerque, New Mexico was responsible for the LOSAT R, but a ground station on Maui, Hawaii controlled the spacecraft. The mission was successful.[38]

We can complete our review of the Cape's DELTA military space operations with a brief look at the NATO IVA mission. As we noted earlier, the NATO IIID communications satellite was launched in November 1984 as a gap filler in the NATO constellation until enhanced NATO IV spacecraft were introduced. The NATO IVA was the first of those enhanced satellites. It was launched from Pad 17B on a Commercial DELTA II Model 7925 vehicle on 8 January 1991. The countdown on January 7th went smoothly for the most part, but a built-in hold had to be extended 68 minutes due to weather constraints. The vehicle lifted off the pad at 0053:01Z on January 8th. The $110,000,000 NATO IVA spacecraft entered its 400 x 19,242-nautical-mile transfer orbit approximately 28 minutes and 48 seconds into the mission. Like its predecessors, the NATO IVA was designed to provide communications between NATO member nations in Europe, the North Atlantic and the eastern seaboard of the United States. The satellite had an operational life expectancy of seven years.[39]

Medium and Light Military Space Operations

ATLAS/CENTAUR Missions at the Cape

ATLAS/CENTAUR military space operations deserve further comment before we move on to light military space operations. Following the FLTSATCOM-E launch on 6 August 1981, there was a lull in military space operations at Complex 36 for about four and one-half years. The Air Force eventually called on NASA to arrange the launch of two more FLTSATCOM spacecraft on General Dynamics' ATLAS G/CENTAUR vehicles in 1986, and the first of those ATLAS Gs arrived at the Skid Strip on 5 March 1986. The booster was destined to launch the FLTSATCOM-G model satellite, which would be redesignated the FLTSATCOM F-7 once it was in orbit. Unfortunately, the mission was delayed several months pending the investigation of the DELTA 178 launch failure, which occurred in May 1986. The FLTSATCOM-G spacecraft finally arrived at the Cape on 29 September 1986. The satellite was checked out at NASA's Hangar AM, then it was moved to Explosive Safe Area 60 (ESA-60) on October 19th. At ESA-60, the spacecraft was mated to its Apogee Kick Motor (AKM), and it was loaded with attitude control propellant. The payload shroud was attached on November 11th, and the spacecraft and fairing were mated to the launch vehicle on 21 November 1986. Prelaunch preparations were completed, and the countdown endured only one unplanned hold of 25 minutes before the count resumed. The ATLAS G/CENTAUR lifted off Pad 36B at 0230:01Z on 5 December 1986. The launch was highly successful, and the spacecraft was injected into the proper 90 x 19,422-nautical-mile transfer orbit. Approximately 48 hours after the launch, ground controllers fired the spacecraft's AKM to circularize the FLTSATCOM's orbit and reduce its inclination to five degrees to the equator. The AKM burn was adjusted to let the FLTSATCOM F-7 "drift" into final position approximately 19,422 miles above the equator.[40]

The other FLTSATCOM mission involved the FLTSATCOM-F spacecraft, which was a slightly shorter version of the FLTSATCOM-G minus EHF communications. The FLTSATCOM-F spacecraft arrived at the Cape on 13 April 1986, and it was scheduled to be launched on an ATLAS G/CENTAUR in December 1986. Unfortunately, the FLTSATCOM-F went into storage after the DELTA 178 launch failure, and it was bumped in the launch schedule by the FLTSATCOM-G mission. Following the launch of the FLTSATCOM-G on December 5th, the FLTSATCOM-F's ATLAS G booster arrived at the Cape on December 9th. The booster was erected at Pad 36B on December 10th, and the CENTAUR upper stage was mated on December 11th. Power-up testing for the ATLAS G/CENTAUR began on 19 December 1986. The spacecraft was taken out of storage in early February 1987. Following testing, the satellite was transported to ESA-60 for its AKM installation and fueling operation in early March. The

FLTSATCOM-F was transported to the launch pad on March 15th, and it was mated to the launch vehicle shortly thereafter. Prelaunch preparations continued, and the countdown was picked up at 1345Z on March 26th. In one of the most disappointing days in the Cape's history, the FLTSATCOM-F was launched through heavy cloud cover on 26 March 1987 only to be struck by lightning and destroyed. NASA's formal mishap investigation concluded that there was "no convincing evidence" that an important criterion-the avoidance of potential electrical hazards-was met by the launch crew. Among the investigation team's recommendations: all directives should be clarified to ensure that they are not ambiguous concerning the duties and responsibilities of launch team members (e.g., weather officers and launch directors).[41]

Figure 116: FLTSATCOM F launch 26 March 1987

As if the FLTSATCOM-F mission failure were not enough, an accident at Pad 36B in July 1987 pushed back the next FLTSATCOM mission indefinitely. The mission (FLTSATCOM F-8) had been scheduled for launch in July, but an oxygen leak was detected in the launch vehicle's CENTAUR interstage adapter during a Terminal Countdown Demonstration Test (TCDT) on 17 June 1987. Another TCDT was performed on July 12th to locate the source of the leak, and a de-mating operation was underway on 13 July 1987 to reach the source of the leak and correct it. Unfortunately, a workstand was pulled off working level 26E as one of the Mobile Service Tower's platforms was retracted from the vehicle. The falling workstand bounced off the lower platform (25E) and struck the CENTAUR's liquid hydrogen tank. The CENTAUR tank ruptured, and the upper part of the launch vehicle dropped and rotated. The CENTAUR was ruined, and it would have to be replaced. In retrospect, this costly accident could have been prevented easily, but fate dictated otherwise. The FLTSATCOM F-8 mission was delayed more than two years as a consequence. NASA planned to transfer sponsorship of the Cape's ATLAS operations to the Air Force following FLTSATCOM F-8's launch, so the accident had an impact on the timeliness of that transfer as well.[42]

Like the FLTSATCOM F-7 spacecraft, the FLTSATCOM F-8 carried UHF, EHF and X-Band communications equipment. The satellite was built by TRW's Defense and Space Systems Group, and it was expected to remain operational in space for ten years. Though satellites were launched toward different locations in the FLTSATCOM constellation, they were all placed in 90 x 19,422-nautical-mile transfer orbits by their launch vehicles. The orbits were circularized about two days into the mission, and the satellites were allowed to drift into position at an altitude of approximately 19,422 nautical miles. The countdown for the F-8 mission got underway at 0012Z on 25 September 1989. The countdown was normal except for a 44-minute extension for weather during the final built-in hold. Once the weather constraint lifted, the countdown resumed, and the ATLAS G/CENTAUR was launched at 0856:02Z on September 25th. The mission was successful.[43]

Figure 117: FLTSATCOM F-8 launch
25 September 1989

The FLTSATCOM F-8 launch was, quite literally, the end of an era at the Cape. NASA did not sponsor any more ATLAS/CENTAUR launch operations after the F-8 mission, and it soon surrendered responsibility for ATLAS/CENTAUR facilities to the Air Force. The NASA/Air Force agreement governing the transfer of ATLAS/CENTAUR program accountability was signed by Air Force Secretary Donald Rice on 29 November 1989 and countersigned by NASA Administrator Richard H. Truly on 22 January 1990. Under the terms of that agreement, Complex 36 and all dedicated ATLAS/CENTAUR facilities at the Cape were transferred to the Air Force at no cost in "as-is" condition. The Air Force assumed accountability for the operation, maintenance and configuration management of Complex 36, both its launch pads, the blockhouse and the facility's ground support equipment. Any future NASA missions launched from Complex 36 would be "accorded the same USAF program consideration as the USAF ATLAS II program," and the two agencies agreed to exchange technical and financial information on ATLAS II boosters carrying NASA payloads. On the commercial side of the house, ESMC and General Dynamics prepared a Joint Operating Procedure (JOP) in March 1990 to outline commercial launch operations responsibilities at Complex 36. Under the JOP, the 6555th Aerospace Test Group's people monitored testing and troubleshooting on both launch pads, but General Dynamics was responsible for fabricating, testing, launching and supporting the Commercial ATLAS launch vehicle. In essence, the company ran its own commercial operation on Pad 36B with safety supervision by Air Force officials. Understandably, the Air Force focused greater attention on Pad 36A because military space missions would be launched from that pad. In practice, the "A" launch pad was for "Air Force" missions and the "B" launch pad was for "business" launch operations.[44]

Medium and Light Military Space Operations

Modification of Cape Facilities for ATLAS II/CENTAUR Operations

With a full manifest of military ATLAS II launches in prospect, General Dynamics set to work preparing the old ATLAS/CENTAUR complex for the new ATLAS II and ATLAS IIA vehicles. Pad 36A needed a new Umbilical Tower (UT), and a major sandblasting and recoating effort was required to halt corrosion on 36B's UT and the Mobile Service Towers (MSTs) on both launch pads. The top of Pad 36A's MST would be cut off just above Level 16, and a 40-foot-tall "splice" consisting of four working levels would be inserted between Level 16 and the MST's old roof. A new elevator would be added to the west side of the MST, and all utilities and operational services would be extended to the new levels. Bridge cranes, traction drive systems and pneumatic work platforms needed repairs. Electrical wiring conduits and water deluge systems needed to be repaired or replaced. Cracks in the blockhouse's roof and walls had to be patched. High pressure storage tanks, safety equipment, propellant pumps and compressors had to be checked and repaired as required. Finally, all propellant systems and storage tanks would have to be recertified to meet new, stricter Air Force safety requirements.[45]

Figure 118: Aerial View of Complex 36
May 1989

On 1 December 1988, Bechtel won the General Dynamics contract to build the Umbilical Tower for Pad 36A. Bechtel began removing old concrete in February 1989, and the tower's foundation support pilings were driven and load-tested by June 1989. Concrete was poured for the base of the UT in June, and the basic structure was completed in October 1989. The new UT was completed around the middle of February 1990, and it was turned over to General Dynamics. In the meantime, primary sandblasting on Pad 36A's MST was completed by mid-January 1990, and work on that tower's 40-foot extension began on 29 January 1990. The basic structure was up by late August, and the extension was virtually complete by mid-September 1990. Gas storage vessel calculations and systems walkdown inspections were completed in December 1990 and February 1991, but General Dynamics admitted that many other systems would not be completed on time. By the end of September 1991, there were still unresolved problems with Pad 36A's bridge crane, MST drive system, MST erection hoist, the east elevator and the launcher. Though a Wet Dress Rehearsal for the site was completed successfully toward the end of October 1991, Quality Assurance officials (ESMC/PQC) still questioned the company's compliance with the corrosion control aspects of its contract with the Air Force. Fortunately, many of the remaining loose ends were resolved over the next five months. On 17 March 1992, Mr. Don Tidwell (Air Force Quality Assurance) signed the certificate of acceptance for Pad 36A, but some exceptions were noted. Among the discrepancies listed, Range Safety took a special interest in Pad 36A's launcher. It had some welding problems. General Dynamics agreed to have additional non-destructive testing done to confirm the integrity of the welds. By early May 1992, Range Safety (45 SPW/SEM) was convinced that there were enough safety factors built into the launcher to make it safe for at least two or three more ATLAS II launches. The company would have the welds inspected thoroughly after each launch, and, if the welds were rejected, the contractor would have to accept the Air Force's ruling and delay the next mission.[46]

Figure 119: Complex 36 Mobile Service Tower under construction February 1990

Figure 120: Complex 36 Towers September 1990

Medium and Light Military Space Operations

ATLAS II/CENTAUR Missions

The first two military ATLAS II/CENTAUR missions were launched from Pad 36A during 1992. We will look at those flights presently, but a few comments concerning the launch vehicles and flight sequences are appropriate at this point. The ATLAS II/CENTAUR vehicle was an improved version of the ATLAS G/CENTAUR. The ATLAS II's performance was increased over the old design, and the ATLAS' fuel tank was lengthened nine feet (e.g., from 73 feet to 82 feet). The ATLAS II's uprated Rocketdyne MA-5A booster and sustainer engines gave the ATLAS II 484,000 pounds of thrust at sea level, and the ATLAS I's vernier engines were replaced with a hydrazine roll control system. The new CENTAUR's tank was three feet longer than the ATLAS G's tank, but the two Pratt & Whitney RL10A-3-3A engines used on both vehicles provided about 16,500 pounds of thrust apiece. Vehicle differences aside, certain flight sequences were common to all ATLAS/CENTAUR flights. Following lift-off, the booster, vernier/hydrazine roll and sustainer engines combined to thrust the vehicle on its way. Between 153 and 171 seconds into the flight, the section containing the two booster engines was jettisoned, but the sustainer engine continued to thrust for approximately two more minutes until that stage's fuel was depleted. The payload fairing was jettisoned during the sustainer burn. Two seconds after sustainer engine cutoff, the CENTAUR separated from the ATLAS/ATLAS II, and the CENTAUR's first main engine burn started about 10 to 12 seconds later. Typically, the first burn placed the CENTAUR and its payload into an elliptical parking orbit anywhere from 80 x 282 nautical miles to 80 x 1239 nautical miles above Earth. The CENTAUR's axial thrusters were brought into play for propellant settling, venting and prestart purposes during a coasting period ranging from 13 to 16 minutes of flight time. The CENTAUR's second burn occurred approximately 24 to 26 minutes into the mission, and it accelerated the CENTAUR and its payload into a highly elliptical transfer orbit from which the spacecraft could be sent into final orbit. Following spacecraft separation, the CENTAUR coasted, turned 90 degrees, and fired its reaction control system to carry itself out of the spacecraft's orbit. A final fuel depletion "blowdown" ended the CENTAUR's fiery performance.[47]

Figure 121: Blockhouse scene-ATLAS II Wet Dress Rehearsal October 1991

Figure 122: ATLAS II/CENTAUR launch 11 February 1992

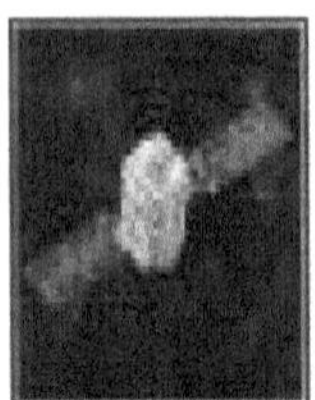

Figure 123: DSCS III Satellite

Figure 124: ATLAS II/CENTAUR erection on Pad 36A

Figure 125: CENTAUR Hoist Operations

Figure 126: Preparation for Tower rollback at Pad 36A

Figure 127: ATLAS II/CENTAUR Pretest

Now we turn to the missions themselves. Following two unsuccessful launch attempts on 6 and 8 February 1992, the first military ATLAS II/CENTAUR mission was launched successfully from Pad 36A on 11 February 1992. The countdown on February 10th was eventful, and two extensions in the launch checklist's built-in holds were required to: 1) repair an environmental control regulator and 2) deal with payload fairing temperature anomalies. The extensions added approximately 70 minutes to the countdown, so the checklist continued through the scheduled 2330Z lift-off on the 10th to the ATLAS II/ CENTAUR's actual lift-off at 0041:02Z on February 11th. The mission marked the first launch of a fully operational DSCS III spacecraft aboard an ATLAS II/CENTAUR vehicle. The CENTAUR's first burn placed the CENTAUR and its payload into an 80 x 282-nautical mile parking orbit, and its second burn accelerated the spacecraft into a highly elliptical 94 x 19,282-nautical-mile transfer orbit. Though

ATLAS and CENTAUR yaw maneuvers contributed to the proper placement of the spacecraft, the first production model Integrated Apogee Boost Subsystem (IABS) was introduced on this mission to change the spacecraft's orbital plane a whopping 26.5 degrees. The spacecraft separated from the CENTAUR approximately three minutes after the CENTAUR's second burn. Two IABS burns were planned, but some fuel migrated after the first burn, and it caused the payload to shift approximately 25 degrees from its spin axis. This made the remaining IABS fuel inaccessible, and the second IABS burn was cancelled. The spacecraft's reaction control thrusters had to be fired over the next several weeks to place the satellite in its proper final orbit. To avoid a reoccurrence of the problem, the second IABS burn was deleted from later DSCS mission profiles, and one full-duration IABS burn was substituted.[48]

Figure 128: Second ATLAS II/CENTAUR launch from Pad 36A 2 July 1992

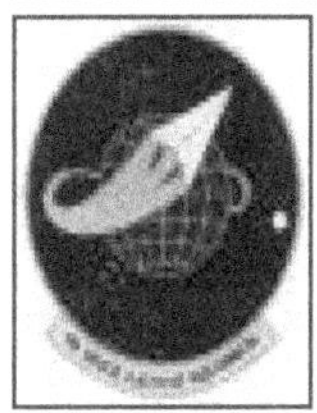

Figure 129: 3rd Space Launch Squadron Emblem

The next military ATLAS II/CENTAUR mission was also a fully operational DSCS III flight. It suffered six launch scrubs in May and June 1992, and the countdown on 2 July 1992 seemed equally unpromising. There were unscheduled holds for weather, a hydrazine leak indication and a Command Message Encoder/Verifier failure, but the launch vehicle lifted off Pad 36A at 2154:01Z as scheduled. The launch was very important to the Air Force's medium launch operations at the Cape. As the second successful military ATLAS II/CENTAUR operation, the mission cleared the way for the activation of the 3rd Space Launch Squadron on 2 July 1992. The activation, in turn, was a highly visible symbol of the Air Force's confidence in the ATLAS II/CENTAUR as an operational launch system. The 3rd Space Launch Squadron looked forward to working with General Dynamics on medium military launch operations at the Cape for many years to come.[49]

Medium and Light Military Space Operations

STARBIRD and RED TIGRESS Operations

There remains the matter of light military space operations at the Cape. Following the demise of the BLUE SCOUT JUNIOR program in the last half of 1965, there were no really light military space launch operations at the Cape until the advent of the STARLAB/STARBIRD program in the mid-to-late 1980s. Under the Strategic Defense Initiative (SDI) program of that period, a series of STARLAB experiments was planned to evaluate laser beam pointing, target acquisition, cooperative beacon tracking and stability. A double neodymium YAG laser would be used to illuminate various targets (e.g., ground targets, coplanar space targets and target boards carried on the fourth stage of a STARBIRD launch vehicle), and a helium-neon laser would be used to score the engagements. The STARLAB payload would be carried into orbit aboard Spacelab in the Space Shuttle's cargo bay. Under the original concept, two or three STARBIRD vehicles would be launched during a Shuttle/STARLAB mission, but only one STARBIRD vehicle would be launched on a "single pass." Wake Island was designated as one of two STARBIRD launch sites, and the Cape was chosen for the other STARBIRD site. Based on the Program Introduction (PI) document distributed on 12 May 1987, the 6550th civil engineers considered Complex 20 to be the best launch site for the STARBIRD program. The U.S. Army Strategic Defense Command (USADC) was responsible for the STARBIRD's design, validation and launch. It also agreed to pay all expenses associated with the STARBIRD program at the Cape.[50]

Figure 130: Complex 20 Blockhouse September 1989

On 23 September 1987, the ESMC Commander signed the Statement of Capability for the STARBIRD program's construction and flight support at the Cape. Though the final configuration of the four-stage STARBIRD vehicle was unknown, the design package for STARBIRD launch facilities was sent to the Army on 13 October 1987 for approval. As program requirements continued to change, the STARLAB Shuttle mission was slipped to September 1990, then November 1990, and eventually August 1991. Though the overall objectives of the program did not change appreciably, technical details-such as the final configuration of the STARBIRD vehicle-changed markedly due to Intermediate Nuclear Force (INF) Treaty constraints. In the meantime, the facility contract for STARBIRD modifications to Complex 20 was awarded to Butler Construction Company on 1 November 1988. The U.S. Army Corps of Engineers met with Butler on 6 December 1988 to discuss: 1) the construction of two STARBIRD launch pads and 2) the renovation of Complex 20's blockhouse and ready building. Butler went to work on Complex 20 in early January 1989 under the supervision of the Army Corps of Engineers. Construction proceeded well, and Complex 20 was turned over to the Range in early December 1989. Earlier, the Eastern Test Range organization had been tasked to design and procure hardware for the retransmission of STARBIRD payload data from the Cape and Wake Island. By April 1989, the ETR's design for an elastic buffer to retransmit the data had been approved. Installation of STARBIRD's major instrumentation systems was expected to begin in February 1990.[51]

Figure 131: Looking East from atop Pad at Complex 20 April 1989

The new STARBIRD facility consisted of two launch pads with 58-foot-tall rail launchers, two Launch Equipment Buildings (LEBs), a Launch Support Center (LSC) and a Payload Assembly Building (PAB). It also had a Payload Support Center (PSC), a Vehicle Support Center (VSC) at the blockhouse and a Missile Assembly Building (MAB) off-site. The facility contractor was still struggling to complete punch list items in several areas in April 1990, but the initial STARBIRD test launch had slipped to early October 1990 by that time. In any event, the first STARLAB mission was not likely to occur before 1992. The first of two STARBIRD rail launchers arrived at Complex 20 on 5 September 1990, and Space Data Corporation erected the launcher (with help from the range contractor) over the next few weeks. Facility construction was virtually complete by the fall of 1990, and the initial STARBIRD launch was rescheduled for mid-December 1990.[52]

Following an untroubled countdown, the first four-stage STARBIRD vehicle was launched from Pad 20A at 0337:25Z on 18 December 1990. The primary objective of the first flight was to provide a booster and post-boost target vehicle for the Army's Ultraviolet Plume Instrument (UVPI) so the latter could demonstrate its ability to acquire, point cameras and track the STARBIRD's third and fourth stages. In addition to the LACE/UVPI spacecraft in near-circular orbit 550 kilometers above Earth, a specially-equipped ARGUS aircraft flew south and west of the Cape to detect the STARBIRD's hardware "in the presence of the (rocket) plume" and capture infrared and visible light data on the flight. A High Altitude Observatory (HALO) aircraft also flew a circuit about 60 kilometers downrange to gather spectrometric data and validate the various stages' plumes. The Innovative Science and Technology Facility (ISTF) on Merritt Island was tasked to cover the STARBIRD's first stage separation and impact, and Malabar's Atlantic Laser Ground Station (ALGS) directed green laser light on the vehicle in a demonstration of passive and active tracking abilities.[53]

The STARBIRD suborbital flight on 18 December 1990 took just 228 seconds to complete, and the vehicle's fourth stage splashed down approximately 262 miles downrange. In that brief space of time, the flight validated the STARBIRD's performance and allowed the Army to characterize the vehicle's rocket plumes. The experiment successfully demonstrated the Ultraviolet Plume Instrument's tracking abilities, and the mission's combined data collection effort helped resolve tracking issues associated with passive sensor algorithms and laser return algorithms. Within hours of the flight, the Strategic Defense Initiative Office announced the STARBIRD mission looked like "a 100 percent success."[54]

No STARBIRD missions were launched from Complex 20 in 1991 or 1992, but the facility supported a different military space effort in 1991. Under the RED TIGRESS program, two ARIES single-stage M56A1 solid rocket boosters were launched from Pad 20A's STARBIRD rail launcher on 20 August and 14 October 1991. Both flights were sponsored by the Foreign Technology Division's Strategic Defense Initiative Office. The purpose of each flight was to boost and eject nine separate experiments into space between 140 and 546 seconds after lift-off. Each flight was expected to last only ten minutes, but considerable effort was required to coordinate each mission's instrumentation requirements and get the flights off the ground. In addition to a full complement of tracking radars, telemetry and optical systems, five specially instrumented aircraft, the HAYSTACK/MILLSTONE and PAVE PAWS radar systems, Malabar's Atlantic Laser Ground Station and Goddard Space Flight Center's large visible optics tracking telescope were pressed into service to gather an enormous amount of spectroscopic and radar data on each flight.[55]

Figure 132: STARBIRD Vehicle on Rail Launcher December 1990

Figure 133: STARBIRD launch 18 December 1990

Figure 134: RED TIGRESS Vehicle on Launcher

Figure 135: RED TIGRESS launch 20 August 1991

The first RED TIGRESS flight was scheduled for 15 August 1991, but the contractor (Orbital Sciences

Corporation) needed more time to test the ARIES' flight termination system. The launch was rescheduled for August 20th. Unfortunately, following lift-off at 0945:48Z, the ARIES veered 90 degrees off-course, and it had to be destroyed about 23 seconds into the flight. The second RED TIGRESS flight should have followed within 72 hours of the first flight, but it was delayed approximately two weeks while the flight failure was investigated. A countdown was attempted on 2 September 1991, but it was scrubbed due to a vehicle telemetry problem. The mission was delayed through the end of September to give the contractor time to handle booster flight control and hardware problems at OSC's facility in Chandler, Arizona. The second ARIES vehicle finally got off the ground at 1017:20Z on 14 October 1991. Though user requirements on two of four telemetry links were not met due to insufficient signal strength late in the mission, Jonathan Dickinson's telemetry performance generally met expectations. Other instrumentation support was highly satisfactory, and the Orbital Sciences Corporation was reportedly "very happy" with the successful mission.[56]

Medium and Light Military Space Operations

U.S. Air Force and NASA Leadership in Space

In Chapter II, we began our discussion of military space operations with some of the largest and most sophisticated space boosters in the world. In Chapter III, we concluded with some of the smallest and simplest suborbital vehicles. Whatever their size and shape, military space operations were -- and are -- an important part of the nation's space effort at Cape Canaveral. As we have seen, some Defense Department missions and space launch vehicles are distinctly military. Others are intertwined with NASA's space initiatives in fundamental ways. While almost nothing in space can be accomplished without the steadfast support of scientists, laboratories, spacecraft companies and launch contractors, it is fair to say that NASA and the Department of Defense have been partners in providing the driving force behind America's space effort from the late 1950s onward.

One of the spin-offs from the U.S. Government's leadership in space was the development of domestic commercial space launch operations. The U.S. commercial space industry grew out of the technological seeds spread by government contracts, and the Defense Department was one of the largest government sponsors for those contracts. While the Air Force and NASA are not allowed to subsidize commercial space efforts at the Cape, unused government facilities may be operated by commercial contractors as long as the government is reimbursed for the direct costs associated with their operation. (Commercial contractors run their own services, but they are at least partially dependent on government facilities for their spacecraft processing and launch operations.) While we may hope the commercial space industry will be able to "go it alone" in the 21st Century, American spacecraft and launch vehicle contractors still look to NASA and the Department of Defense to sponsor new directions in space vehicle and launch facility architecture in the waning years of the 20th Century.

It is logical, therefore, to pay serious attention to the government's long-term proposals for new space boosters and launch infrastructures even if the concepts seem somewhat grandiose and impractical in the short term. In the final chapter, we will look at government-sponsored space vehicle and launch facility studies in the mid-to-late 1980s and 1990 and the fiscal and political realities that toned down their ultimate effect. The reader is cautioned: the studies describe space operations at the Cape only as they *might* be. They should not be construed as an endorsed blueprint for the future or a coordinated plan. At best, they may be thought of as a direction-a new way military and civilian space operations could be launched from the Cape in the 21st Century, based on emerging technologies rooted in the 20th.

Chapter Three Footnotes

Lt. Colonel Bobby J. Hilbert
Lt. Colonel William J. Sparkman succeeded Lt. Colonel Hilbert as ATLAS Systems Division Chief at the end of September 1971, and he continued in that capacity until April 1973.
Lt. Colonel Warren G. Green assumed the Chief's duties following Sparkman's departure in April, and he remained in that post until he became Chief of the Space Launch Vehicle Systems Division in November 1975.

Complex 13
General Dynamics had 90 people assigned to its operations at Complex 13 in January 1971, and Lockheed had 161 people on the job. The numbers fluctuated between 70 and 100 personnel for GDC over the next five years, but Lockheed's numbers continued on a fairly steady downward trend until the company instituted its factory-augmented launch team concept in the early part of 1974. Lockheed's launch base strength quickly dropped from 117 to 70 personnel following that change, and the Lockheed workforce finally leveled off at about 60 people by the middle of 1975. General Electric, Burroughs and Rocketdyne maintained virtually the same numbers of employees throughout the period (e.g., 25, 10 and two people in 1971 and 24, 10 and two people in 1975).

payloads branch
Captain Frederick R. Wohrman was Chief of the Payloads Branch, and he continued as the Manager for TITAN IIIC Satellite Systems in 1973. Major Jerry H. Freer succeeded Major Wohrman in the latter half of 1973, and Freer continued as Chief of the Space Satellite Systems Launch Operations Branch before he became Chief of the newly created Satellite Systems Division in November 1975.

6595th Commander's request to reorganize
In a letter to the 6555 Test Group's commander in June 1975, Colonel William C. Chambers (6595th Aerospace Test Wing Commander) requested the reorganization of the 6555 Test Group's ATLAS and TITAN payload sections into one new division to make the most efficient use of "limited Wing/Group manpower." Colonel John C. Bricker received approval to initiate the reorganization in October 1975, and the effective date of the change was established as 1 November 1975. Under the new setup, the Satellite Systems Division was authorized six officers, seven airmen and three civilians. The Space Launch Vehicle Systems Division received slots for 20 officers, 31 airmen and 13 civilians. The Space Transportation System Division was authorized six officers and one civilian, and the Resources Management Division was given slots for one officer, seven airmen and one civilian. Colonel Bricker's immediate office staff consisted of one officer and two civilians.

NATO IIB

The NATO IIB weighed approximately 535 pounds at launch and 285 pounds on orbit. Cylindrical in shape, the spacecraft was 54 inches in diameter and 32 inches high. (Its antenna assembly added another 31 inches to the length of the satellite when deployed.) The satellite was placed in a 22,000-mile-high geosynchronous orbit midway between Africa and South America in line with the equator. The satellite joined the NATO IIA communications satellite, which had been launched successfully in March 1970. The Kennedy Space Center's Unmanned Launch Operations Directorate was contracted to launch the NATO IIB for the Air Force, which acted as NATO's agent. The NATO IIB linked NATO Headquarters in Brussels, Belgium with national capitals and NATO command locations on land and sea.

ATLAS/AGENA

For a typical mission, the ATLAS booster/sustainer stage was transported from GDC's plant to the Cape and received at Hangar J. Following its receiving inspection and nacelle installation, the booster/sustainer was erected about ten weeks before launch. The AGENA upper stage was flown from Sunnyvale, California to the Cape, and it was received and inspected in Hangar E. Following mechanical checks, the AGENA was transferred to the launch pad about three weeks before launch. The payload arrived about two to three weeks before launch, and it was taken directly to the pad for mating and prelaunch tests. Compatibility checks and launch readiness tests were performed during the last two weeks before the mission.

installation and system validation

After AFSCF checks were completed, the New Hampshire Remote Tracking Station performed an operational simulation for the RVCF using a telemetry and command simulator. Put simply, the tracking station sent computer commands to the RVCF via commercially leased data circuits, and the RVCF formatted and transmitted command tones to the simulator. The RVCF subsequently received the simulator's downlink telemetry and retransmitted it to the Satellite Test Center in Sunnyvale, California for printout.

space-related programs

The programs included DSCS II, the Defense Support Program, the NATO III, the Universal Payload Fairing, Space Test Program 74-1, and the Applications Technology Satellite F (ATS-F) program. The FLTSATCOM program may also be included, though it was transferred to the ATLAS Systems Division in June 1973. In connection with this list, only the NATO III, SKYNET and FLTSATCOM programs will be covered in this chapter.

SKYNET satellites

SKYNET A was launched from Pad 17A on 21 November 1969, and it was placed in a near synchronous orbit above the equator over the Indian Ocean. It was the first spacecraft in a planned constellation of one operational communications satellite and one standby satellite. SKYNET was designed to provide the British Ministry of Defence with secure communications between Great Britain and posts as far away as Singapore. Cylindrical in shape, the SKYNET A was 32 inches in diameter and 54 inches long. With its antennas extended, its overall length was about 62 inches. The spacecraft weighed 535 pounds at launch and 285 pounds on orbit. The second satellite, SKYNET B was the same size and weight as the SKYNET A, but its launch was not as fortunate: following its lift-off from Pad 17A on 19 August 1970, the SKYNET B was lost coming out of its transfer orbit after its apogee kick motor misfired midway through its intended 27-second burn.

SKYNET II-A
The SKYNET II series satellites were cylinders 75 inches in diameter and 82 inches high. Each weighed 960 pounds at lift-off and had an operational life expectancy of five years.

NATO IIIA communications satellite
The NATO III series satellites were built by Aeroneutronic Ford Corporation's Western Development Laboratories Division under a Space and Missile Systems Organization contract for NATO. The NATO III network was planned as a constellation of three satellites, all launched by NASA for the Air Force at Cape Canaveral. The main body of each satellite was a cylinder 86 inches in diameter and 88 inches long. (Protruding antennas gave the spacecraft an overall length of 122 inches.) The NATO III spacecraft weighed 1,543 pounds at lift-off and approximately 700 pounds (following Apogee Kick Motor fuel depletion) on orbit. The satellite's surface was covered with solar cells to provide electrical power, and the spacecraft was equipped with two wide beam transponders and one narrow beam transponder to provide secure communications to the NATO member nations. The satellite had an operational life expectancy of seven years.

DELTA booster
The booster consisted of an extended long-tank THOR first stage, TRW second stage, Thiokol third stage and nine Castor II solid rocket motors. A DELTA Inertial Guidance System (DIGS) in the second stage of the vehicle guided the first two stages to the release point for the third stage and payload. A timer in the third stage initiated the spacecraft separation.

spacecraft system performed
The catalyst-bed thruster heater did not perform as well as expected, but it had no effect on the mission's success.

Lt. Colonel Russell E. Vreeland, Jr.
Colonel Vreeland succeeded Major Freer as Satellite Systems Division Chief following Freer's reassignment as Chief, Resources Management Division in the last half of 1976. Colonel Vreeland continued as Division Chief at least through September 1978. He was succeeded by Colonel Charles H. MacGregor sometime in FY 1979.

FLTSATCOM spacecraft
The FLTSATCOM spacecraft consisted of two hexagonal modules equipped with antennas and a solar array. In its deployed configuration, the satellite was approximately 24 feet long and 44 feet wide (across its solar panels). It had an operational life expectancy of five years.

super high frequency communications
The Naval Electronics System Command provided overall program management for the FLTSATCOM, but the Air Force was given 12 narrow-band communications channels for use in the Air Force Satellite Communications System.

SKYNET IV
The British Ministry of Defence initially supported a SKYNET III follow-on to the SKYNET II program, but it eventually cancelled that project in favor of the SKYNET IV. Unlike its American-built predecessors, SKYNET IV communications satellites were built by the British Aerospace Dynamics Group.

British
The British expressed an interest in manifesting the SKYNET IVs on all-military Shuttle missions, but the Ministry of Defence was willing to settle for mixed military/commercial flights.

Greenwich Mean Time
All times will be expressed in "Z" (Greeenwich Mean Time) for the remainder of this chapter.

ARIANE 4 launch vehicle
The ARIANE 4 used for the mission was equipped with two strap-on liquid rocket engines and two strap-on solid rocket motors to lift a dual payload consisting of the ASTRA-1 communications satellite and the SKYNET IV-B. That vehicle could lift approximately 3,700 kilograms into geostationary orbit.

first launch attempt
The first attempt was scrubbed due to clouds and lightning in the launch area. Four more attempts were scrubbed for upper level wind constraint violations, and the other two attempts were scrubbed for temperature constraint violations.

military navigation
The Navy sponsored two navigation satellite programs to enhance two-dimensional navigation (e. g., TRANSIT and TIMATION). The Air Force underwrote concept and system design studies for a highly accurate three-dimensional navigation system called System 621B. The System 621B concept was verified in tests and experiments at Holloman AFB, New Mexico and the White Sands Missile Range.

Block I NAVSTAR satellites
The satellites were inserted into orbit by a stage vehicle system developed by Fairchild Space and Electronics Company of Germantown, Maryland. The system employed a spin-stabilized, tandem pair of solid rocket motors (mounted atop the ATLAS F) to boost each 1,720-pound GPS satellite into orbit.

GPS development phase

During the developmental phase, the fledgling Block I GPS constellation was supported with replenishment satellites and the first Block II GPS satellite (designated GPS 12) was built. During the production phase, a full constellation of Block II production satellites would be built and deployed, and large quantities of user equipment would be manufactured and issued to forces in the field.

payload assist modules

This was not McDonnell's first payload assist module contract for the GPS program. As far back as 1978, the Air Force was aware that Block II NAVSTAR satellites would be 200-400 pounds heavier than Block I spacecraft, and Fairchild's original stage vehicle system would not be able to handle the heavier payloads. Consequently, the Space and Missile Systems Organization advertised for a more powerful stage vehicle in October 1978, and it issued a formal Request For Proposal (RFP) on 25 January 1979 for two Space Guidance System II (SGS II) upper stages with an option to deliver and launch five more vehicles by July 1983. McDonnell Douglas was the only contractor to respond to the RFP by the closing date (13 March 1979), and the company was awarded the initial SGS II contract on 14 June 1980. McDonnell Douglas experienced nozzle defects and stability problems with the SGS II's Thiokol Star 48 solid rocket motors, but, with the Aerospace Corporation's help, the contractor resolved its difficulties in 1983. ATLAS E launch vehicles equipped with SGS II upper stages were used to boost the NAVSTAR 9, 10 and 11 satellites into their transfer orbits in 1984 and 1985. Concerning McDonnell Douglas' later contract, the PAM-DII was based on the PAM-D payload assist module already in service on the Space Shuttle. Like the PAM-D, the PAM-DII was designed to boost its payload into an elliptical transfer orbit after the spacecraft was deployed from the Shuttle's cargo bay.

store Block II satellites

Slowing down the production line proved to be a cheaper solution than storing the NAVSTAR II satellites. So, in December 1986, the Air Force asked Rockwell to submit a proposal to extend the production schedule by three years. In May 1988, a change to the contract incorporated the three-year extension at a cost of approximately $96,500,000. The first production NAVSTAR II satellite (GPS-13, a.k.a., NAVSTAR II-1) was delivered to the Air Force in April 1987. Four other production NAVSTAR II satellites were delivered to the Cape by the middle of July 1988.

a newly configured PAM-D

The use of the PAM-D designation is a bit confusing, since an earlier PAM-D had been used with the Space Shuttle before the PAM-DII contract was awarded back in 1984. The new PAM-D developed about 25,000 pounds of thrust. It was intended for use *only* with the DELTA II booster and NAVSTAR II satellite. Thereafter, the PAM-S was designated for use with the Space Shuttle.

NASA

NASA processed requests for DELTA launch services, but the Air Force was going to replace NASA as the government agency responsible for DELTA operations at the Cape. This would occur as soon as Complex 17, Area 57 and most of Hangar M could be transferred to the Air Force-presumably, after the DELTA 181 mission was launched on 8 February 1988. In accordance with an Air Force/NASA agreement signed by Secretary Aldridge and NASA Administrator James C. Fletcher on 1 July 1988, the KSC Director signed a Memorandum of Understanding (MOU) between KSC and ESMC authorizing the transfer of DELTA launch site operations to the Air Force. The MOU was signed on 16 August 1988. Though NASA continued to process the DELTA 183 payload at KSC, transfer of the Cape's DELTA-related facilities was to begin immediately. The 6550th's civil engineers received and distributed NASA's accountability transfer forms before the end of August, and the signed forms were hand-delivered to NASA's Air Force Management Office on 29 September 1988. Complex 17's real property transfer was essentially complete by 10 October 1988.

extended their workdays
McDonnell Douglas maintained that schedule for about a year before easing back to 60-hour workweeks in late August 1989. The Test Group's Medium Launch Vehicle Division maintained a six-day, 72-hour workweek from July 1988 through mid-August 1989.

Block IIA NAVSTAR satellites and Hercules Graphite Epoxy Motors (GEMs)
The Block IIA satellites were approximately 400 pounds heavier than their Block II predecessors because of extensive mechanical differences between the two series of satellites. To handle the additional payload, the Rocketdyne main engine nozzle was increased from an 8:1 expansion ratio to a 12:1 expansion ratio, and the Morton Thiokol Castor IVAs were replaced with more powerful Hercules Graphite Epoxy Motors (GEMs). The Hercules GEMs were about six feet longer and 3,000 pounds heavier than the Castor IVAs (e.g., 401.6 inches vs. 323.4 inches and 28,657 pounds vs. 25,562 pounds), but the motor's lighter graphite epoxy casing allowed all the additional weight to be translated into fuel. Each GEM generated 144,533 pounds of maximum thrust-approximately 25,000 pounds more than the Castor IVA's maximum-and the GEM's total impulse power was about 16 percent higher than the Castor IVA.

second stage's engine ceased firing
The Model 6925's second stage engine cutoff occurred approximately 11 minutes and 29 seconds after launch. The same event occurred about a minute earlier for the Model 7925.

transfer orbit
Only the Model 7925 used the second stage engine to assist with the transfer orbit maneuver. The second stage's burn lasted approximately 20 seconds, and the second stage coasted before separating from the vehicle about 21 minutes and 2 seconds into the flight. The Model 7925 fired its third stage motor about 21 minutes and 40 seconds after lift-off, and it inserted the spacecraft into the proper transfer orbit. There was no second burn for the Model 6925's second stage. The second stage merely separated from the vehicle after coasting in low-Earth orbit for about nine and a half minutes. The Model 6925's third stage ignited about 21 minutes and 33 seconds after lift-off, and it pushed the payload into its proper transfer orbit.

five earlier countdowns were attempted
Countdowns were attempted on May 20, May 21st, May 23rd, May 24th and 9 June 1989 before

the NAVSTAR II-2 mission was launched on June 10th. A Liquid Oxygen (LOX) valve problem caused the launch scrub on May 24th, and the other launch attempts were scrubbed for weather.

LOSAT-X payload
The LOSAT-X should have been launched along with two other Strategic Defense Initiative payloads (e.g., LOSAT L and LOSAT R) on a Commercial DELTA II launch vehicle on 14 February 1990. Unfortunately, the "X" payload was not ready in time for that launch, so it was remanifested as part of the NAVSTAR II-11 flight. The LOSAT-X was mounted sideways on the DELTA II's second stage, and it was jettisoned approximately one hour, five minutes and thirty seconds into the flight-about 44 minutes and 27 seconds after second stage/third stage separation. A ground station in Boulder, Colorado provided primary control for the LOSAT-X on its classified mission. The Consolidated Space Test Center at Sunnyvale, California provided a backup control capability.

three separate countdowns
The first two countdowns were aborted on 4 and 5 April 1992 due to upper level winds. Weather was also a concern during the night of April 9th, but wind conditions improved. The third and final countdown was uneventful, and the vehicle lifted off the pad at 0319:59.988Z on April 10th.

vehicle misfire
The DELTA II vehicle was placed in a safe condition after the misfire, and it took two hours to detank the kerosene and liquid oxygen propellants from the vehicle's first stage.

impact
One of the seven-NAVSTAR 4-was rated marginal due to subsystem failures, and it needed extensive support from ground controllers to remain in operation during 1989. It was finally removed from the operational GPS constellation in August 1989.

six more NAVSTAR II satellites
The GPS constellation consisted of six Block I and three Block II satellites before six more Block II spacecraft were added to the constellation.

four Strategic Defense Initiative (SDI) missions
Admittedly, the first two SDI missions were launched on NASA-sponsored vehicles, and NASA's contract with McDonnell-Douglas remained in effect until the third SDI payload (DELTA STAR) was processed. The fourth SDI mission was launched on a Commercial DELTA II vehicle. Launch vehicles aside, the payloads and missions were military.

countdown for the DELTA 181 mission
The countdown marked the second launch attempt for the DELTA 181 mission. The first launch attempt was scrubbed at 2219Z on 4 February 1988 due to a second stage fuel tank valve malfunction.

LIDS
The LIDS was placed onboard to measure energy over a wide optical spectrum.

ATLAS G/CENTAUR vehicles
The ATLAS G/CENTAUR vehicle was 137 feet 7 inches long. Its booster/sustainer assembly generated a total thrust of 437,500 pounds of thrust at lift-off, and the CENTAUR upper stage provided 33,000 pounds of thrust. The gross weight of the vehicle-minus spacecraft-was 356,120 pounds.

FLTSATCOM-G
The FLTSATCOM-G weighed about 4,977 pounds on the ground and 2,693 pounds in geosynchronous orbit after its apogee kick motor propellants had been expended. The spacecraft was seven feet six inches wide and 22 feet 10 inches long. Its deployed solar array measured 43 feet five inches. The satellite was equipped with standard UHF and X-Band communications, and it also carried an experimental module for extremely high frequency (EHF) communications. Though the EHF module only had a test life expectancy of two years, the UHF suite of communications had an operational life expectancy of at least ten years. The FLTSATCOM-G was built by TRW's Defense and Space Systems Group.

the mission was delayed several months
Due to similarities in the ATLAS and DELTA main engineering electronics relay boxes and wiring harnesses, the FLTSATCOM F-7 launch was slipped from 22 May 1986 to 5 December 1986.

struck by lightning and destroyed
Following the vehicle's lift-off at 2122:00.768Z on the 26th, four strokes of lightning were recorded by two 16 mm cameras, a NASA video van and local TV station equipment covering the launch. Physical evidence and other information revealed that the fourth stroke of lightning struck the ATLAS G/CENTAUR just before the vehicle's computer issued a positive yaw command. The vehicle began to heel over about a tenth of a second later (e.g., at T plus 48.46 seconds), and the vehicle started to break up at T plus 50.7 seconds. The spacecraft separated from the vehicle at T plus 50.96 seconds. Subsequently, destruct commands were sent, but they proved fruitless. The flight ended as a result of vehicle break-up.

workstand was pulled off
The workstand had been fouled with an improperly plugged overhead light power cord. As the platform retracted, the cord pulled taut, and the line tipped the workstand over the edge of the platform. The cord broke loose, releasing the workstand to tumble on the platform below.

facilities
The facilities included Complex 36, Hangar J, Hangar K, Paint Oil Lockers J and K, Storage Building 55010, Barrel Storage Area 55040, Storage Building 1737, Little J (Building 55000) and Little K (Building 55001). All ground support equipment, space test equipment and spare parts had already been transferred to General Dynamics Space Systems Division (GDSS) via a "trade and barter" agreement in the CRRES/AC-69 contract.

NASA payloads
Under the agreement's implementation plan, signed by the KSC Director and ESMC Commander in late January 1990, NASA did not intend to participate in day-to-day launch vehicle preparations. NASA would be apprised of work schedules, test results, crew certification, etc., and a senior management assessment team would be present during the launch countdown. While a NASA Mission Director would be present to provide the go/no-go decision on launch day, the Air Force would direct the launch contractor for any NASA payloads launched on Air Force-procured vehicles.

launch pads
In the interest of clarity, we will examine only the principal changes to Pad 36A. Pad 36B was inherently a commercial launch facility, and its refurbishment had very little impact on military space operations at Complex 36.

certificate of acceptance for Pad 36A
The first military ATLAS II/CENTAUR mission was launched by General Dynamics from Pad 36A on 11 February 1992. The launch occurred more than a month before the government's acceptance certificate for the launch site was signed.

first two military ATLAS II/CENTAUR missions
The military missions were not the first flights for the ATLAS II at the Cape. The first Commercial ATLAS II/CENTAUR was launched from Pad 36B on 7 December 1991. It carried the EUTELSAT II, a European telecommunications satellite, into a 455 x 22,216-nautical-mile transfer orbit on that date.

Rocketdyne MA-5A booster and sustainer engines
The ATLAS G booster/sustainer provided a combined thrust of 437,500 pounds of thrust at sea level. The booster's two Rocketdyne MA-5 engines provided 377,500 pounds of that total power, and the sustainer's single MA-5 provided the remaining 60,000 pounds of thrust. The ATLAS II booster/sustainer was configured similarly, but with the increased thrust noted.

new CENTAUR's tank
A more improved vehicle, the ATLAS IIA/CENTAUR, had a CENTAUR upper stage equipped with two Pratt & Whitney RL10A-4 engines rated at 20,800 pounds of thrust each. The first ATLAS IIA/CENTAUR was launched on a commercial flight from the Cape on 10 June 1992.

two unsuccessful launch attempts
The launch attempt on February 6th was scrubbed due to excessive winds. A second launch attempt on the 8th was scrubbed due to excessively low tanking temperatures noted during the CENTAUR's upper stage fueling sequence.

DSCS III
Three earlier DSCS III spacecraft had been orbited for developmental purposes, and two operational DCSC IIIs had been orbited via the Shuttle. The satellite launched on 11 February 1992 was deployed as an operational communications relay between command posts and U.S. military forces in the field.

activation of the 3rd Space Launch Squadron on 2 July 1992
The squadron's activation was announced to have taken place on 2 July, but the activation ceremony was actually held on 3 August 1992.

the first four-stage STARBIRD vehicle
The STARBIRD booster consisted of a TALOS first stage, a SERGEANT second stage and two ORBUS 1 upper stages. The 57-foot-long vehicle was guided with inertial sensors and a microprocessor. Fixed fins were used to control the first stage, and ailerons were used on the second stage. Air and jet vanes controlled the vehicle during its second stage burn. Third and fourth stage pitch and yaw were accomplished by a gimbaled nozzle, but their roll was controlled by a cold gas Attitude Control System (ACS) housed in the fourth stage. A cold gas control system was used to stabilize the payload scoreboard in relation to the LACE/UVPI satellite used in the experiment. The STARBIRD vehicle was built by the Space Data Division of the Orbital Sciences Corporation.

full complement of tracking radars, telemetry and optical systems
A total of five radars on Merritt Island, the Cape, Patrick and Jonathan Dickinson Missile Tracking Annex were tasked to track the RED TIGRESS vehicle. Two surveillance radars also covered the launch danger area for range safety purposes. The Eastern Range's optical support for RED TIGRESS flights (e.g., Contraves units, Instrumented Ground Optical Recorders, Distant Object Attitude Measurement Systems, Intermediate Focal Length Optical Trackers and ITEK units) rivaled the coverage given any major launch operation. The Cape provided command/destruct capabilities, and TEL-IV and Jonathan Dickinson handled telemetry.

ARIES
A failure analysis revealed that the ARIES' onboard computer had been loaded with ground test software instead of the vehicle's flight program. This "operator oversight" ruined the mission.

Chapter Three Endnotes

1. 6555th Aerospace Test Group History, 1 January - 30 June 1971, ATLAS Systems Division Historical Section, pp. 3, 4; 6555th Aerospace Test Group History, 1 July - 31 December 1971, ATLAS Systems Division Historical Section, pp. 3, 4; 6555th Aerospace Test Group History, 1 January - 30 June 1972, ATLAS Systems Division Historical Section, p. 4; Briefing Slides, ATLAS Systems Division, "Contractor Support" and "Facilities," o/a January 1972; 6555th Aerospace Test Group History, 1 July - 31 December 1972, ATLAS Systems Division Historical Section, p. 4; 6555th Aerospace Test Group History, 1 January - 30 June 1973, ATLAS Systems Division Historical Section, pp. 3, 4 and TITAN III Systems Division Historical Section, pp. 8, 11; 6555th Aerospace Test Group History, 1 July - 31 December 1973, ATLAS Systems Division Historical Section, p. 4 and TITAN III Systems Division Historical Section, p. 4; 6555th Aerospace Test Group History, 1 January - 30 June 1974, ATLAS Systems Division Historical Section, p. 4; 6555th Aerospace Test Group History, 1 July - 31 December 1974, ATLAS Systems Division Historical Section, p. 3; 6555th Aerospace Test Group History, 1 January - 30 June 1975, ATLAS Systems Division Historical Section, p. 4; 6555th Aerospace Test Group History, 1 July - 31 December 1975, Space Launch Vehicle Systems Division Historical Section, pp. 1, 6 and Satellite Systems Division Historical Section, pp. 1, 3; Letter, 6595th ATW/CC to 6555th ASTG/CC, "Organizational Changes," 25 June 1975; Letter, 6555th ASTG/CC to (unlisted agencies), "Reorganization of the 6555th Aerospace Test Group," 17 October 1975; Letter, MET 27 to SAMTEC/CC, "Organizational Realignment of 6555th Aerospace Test Group (6595ATW/CC ltr, 17 Oct 75)," 13 November 1975.

2. News Release, AFETR Office of Information, "Air Force Eastern Test Range Prepares for NATO-B Launch," 10 January 1971; News Release, NASA/KSC, "Photo No. 102-KSC-71P-176," 2 February 1971; 6555th Aerospace Test Group History, 1 January - 30 June 1971, TITAN III Systems Division Historical Section, p. 8 and ATLAS Systems Division Historical Section, p. 6; 6555th Aerospace Test Group History, 1 July - 31 December 1971, ATLAS Systems Division Historical Section, p. 6; Briefing Slide, ATLAS Systems Division, "Factory to Pad," o/a January 1972; Marven R. Whipple, "Eastern Test Range Index of Missile Launchings, July 1970 - June 1971," undated, p. 31; Marven R. Whipple, "Eastern Test Range Index of Missile Launchings, July 1971 - June 1972," undated, p. 3 (information used is unclassified); Marven R. Whipple, "Eastern Test Range Index of Missile Launchings, July 1972 - June 1973," o/a 10 August 1973, p. 3; 6555th Aerospace Test Group History, 1 January - 30 June 1972, TITAN III Systems Division Historical Section, p. 11; 6555th Aerospace Test Group History, 1 July - 31 December 1972, TITAN III Systems Division Historical Section, pp. 10, 11.

3. 6555th Aerospace Test Group History, 1 January - 30 June 1973, TITAN III Systems Division

Historical Section, pp. 8, 10; 6555th Aerospace Test Group History, 1 July - 31 December 1973, TITAN III Systems Division Historical Section, p. 12; Marven R. Whipple, "Eastern Test Range Index of Missile Launchings, July 1973 - June 1974," undated, pp. 5, 31; 6555th Aerospace Test Group History, 1 January - 30 June 1974, TITAN III Systems Division Historical Section, p. 11; AFETR History, FY 1974, Volume I, Part 2, p. 407; Marven R. Whipple, "Eastern Test Range Index of Missile Launchings, July 1969 - June 1970," undated, p. 42; Whipple, "Index, July 1970 - June 1971," undated, p. 35.

4. 6555th Aerospace Test Group History, 1 January - 30 June 1974, TITAN III Systems Division Historical Section, p. 12; 6555th Aerospace Test Group History, 1 July - 31 December 1974, TITAN III Systems Division Historical Section, p. 13; Marven R. Whipple, "Eastern Test Range Index of Missile Launchings, July 1974 - June 1975," undated, pp. 8, 35; Article, "Skynet Achieves Orbit," *Florida Today*, 25 November 1974; Article, "Skynet IIB Satellite Blasts Off From Cape," *Sentinel Star*, 23 November 1974; ESMC History, 1 October 1984 - 30 September 1986, Volume I, p. 330.

5. 6555th Aerospace Test Group History, 1 January - 30 June 1973, ATLAS Systems Division Historical Section, p. 6; 6555th Aerospace Test Group History, 1 July - 31 December 1974, ATLAS Systems Division Historical Section, p. 6; 6555th Aerospace Test Group History, 1 January - 30 June 1975, ATLAS Systems Division Historical Section, 1 July - 31 December 1975, Satellite Systems Division Historical Section, p. 9; Whipple, "Index, July 1974 - June 1975," undated, p. 3; Marven R. Whipple, "Eastern Test Range Index of Missile Launchings, July 1975 - June 1976," undated, p. ii; Marven R. Whipple, "Eastern Test Range Index of Missile Launchings, July - December 1976," undated, p. ii; Marven R. Whipple, "Eastern Test Range Index of Missile Launchings, CY 1977," undated, p. 3 (information used is unclassified); Marven R. Whipple, "Eastern Test Range Index of Missile Launchings, CY 1978," undated, p. 3 (information used is unclassified).

6. 6555th Aerospace Test Group History, 1 July - 31 December 1975, Satellite Systems Division Historical Section, p. 5 and TITAN III Systems Division Historical Section, p. 11; 6555th Aerospace Test Group History, 1 January - 30 June 1976, Satellite Systems Division Historical Section, pp. 4, 5; Whipple, "Index, July 1975 - June 1976," p. 42; Whipple, "Index, CY 1977," p. 53; Whipple, "Index, CY 1978," pp. 26, 49.

7. 6555th Aerospace Test Group History, 1 July - 31 December 1976, Satellite Systems Division Historical Section, pp. 4, 5; 6555th Aerospace Test Group History, 1 January - 30 June 1977, Satellite Systems Division Historical Section, p. 4; Whipple, "Index, CY 1977," p. 24; Whipple, "Index, CY 1978," p. 49; 6555th Aerospace Test Group History, January - September 1978, p. 9; 6555th Aerospace Test Group History, October 1978 - September 1979, Volume I, p. 22.

8. 6555th Aerospace Test Group History, 1 July - 31 December 1977, Satellite Systems Division Historical Section, pp. 5, 6; 6555th Aerospace Test Group History, January - September 1978, pp.

28, 29; Whipple, "Index, CY 1978," pp. 6, 42.

9. 6555th Aerospace Test Group History, January - September 1978, p. 25; 6555th Aerospace Test Group History, October 1978 - September 1979,, Volume I, pp. 32; Whipple, "Index, CY 1978," pp. 26, 49.

10. 6555th Aerospace Test Group History, October 1978 - September 1979, Volume I, pp. 47, 48.

11. 6555th Aerospace Test Group History Submission for FY 1979, Satellite Systems Division Historical Section, p. 7; ESMC History, 1 October 1979 - 30 September 1981, Volume I, p. 480.

12. ESMC History, 1 October 1979 - 30 September 1981, Volume I, pp. 484, 491.

13. Letter, 6555 ASTG/SIE to 6555 ASTG/CC, "6555 ASTG Historical Report (Oct 82 - Mar 83)," 22 June 1983; Letter, 6555 ASTG/SI to 6555 ASTG/CC, "6555th ASTG Historical Report (April - September 1983)," 28 October 1983; 6555th Aerospace Test Group Historical Report, 1 October 1983 - 31 March 1984, Spacecraft Division Historical Section; ESMC History, 1 October 1984 - 30 September 1986, Volume I, p. 330.

14. Letter, 6555th ASTG/SI to 6555th ASTG/CC, "6555th ASTG Historical Report (April - September 1983)," 28 October 1983; 6555th Aerospace Test Group History, 1 October 1985 - 31 March 1985, Spacecraft Division Historical Section, "NATO IIID;" ESMC History, 1 October 1984 - 30 September 1986, Volume I, pp. 280, 281.

15. 6555th Aerospace Test Group History, April - September 1984, Spacecraft Division Historical Section, "SKYNET;" ESMC History, 1 October 1984 - 30 September 1986, Volume I, pp. 330, 331. ESMC History, 1 October 1988 - 30 September 1989, Volume I, pp. 373, 374; ESMC History, 1 October 1989 - 30 September 1990, Volume I, pp. 127, 294, 295, 323.

16. AFSC History, 1 October 1979 - 30 September 1980, Volume I, pp. 531, 533, 534; Geiger "Launch Summary," 15 December 1992, pp. 83, 84, 85, 87; SAMSO History, 1 January - 31 December 1978, Volume I, p. 63; SAMTO and WSMC History, 1 October 1979 - 30 September 1980, Volume I, pp. 52 and 54; Space Division History, 1 October 1980 - 30 September 1981, Volume I, pp. 246, 247; Space Division History, 1 October 1983 - 30 September 1984, Volume I, p. 180.

17. Space Division History, 1 October 1981 - 30 September 1982, Volume I, pp. 192, 193; Space Division History, 1 October 1982 - 30 September 1983, Volume I, pp. 176, 178; Space Division History, 1 October 1983 - 30 September 1984, Volume I, p. 186; SAMTO and WSMC History, 1 October 1985 - 30 September 1986, Volume I, p. 66; Geiger, "Launch Summary," 15 December 1992, pp. 89, 91, 92, 93, 94.

18. AFSC History, 1 October 1981 - 30 September 1982, Volume I, p. 162; AFSC History, 1 October 1982 - 30 September 1983, Volume I, pp. 226, 413; SAMSO History, 1 January - 31 December 1978, Volume I, pp. 63, 64. SAMSO History, 1 January - 31 December 1979, Volume I, pp. 63, 64; Space Division History, 1 October 1981 - 30 September 1982, Volume I, pp. 193, 194; Space Division History, 1 October 1982 - 30 September 1983, Volume I, p. 183, 184; Space Division History, 1 October 1983 - 30 September 1984 Volume I, pp. 191, 192, 193; Letter, 6555th ASTG/SI to 6555th ASTG/CC, "6555th ASTG Historical Report (April - September 1983)," 28 October 1983; 6555th Aerospace Test Group History, April - September 1984, Spacecraft Division Historical Section, "NAVSTAR Global Positioning System (GPS);" 6555th Aerospace Test Group History, 1 October 1984 - 31 March 1985, Spacecraft Division Historical Section, "NAVSTAR Global Positioning System (GPS);" ESMC History, 1 October 1982 - 30 September 1984, Volume I, pp. 57, 58; Space Division History, 1 October 1984 - 30 September 1985, Volume I, p. 262.

19. Space Division History, 1 October 1984 - 30 September 1985, Volume I, pp. 258, 259; Space Division History, 1 October 1985 - 30 September 1986, Volume I, pp. 250, 251, 257.

20. Space Division History, October 1985 - September 1986, Volume I, pp. 252, 253, 254; Space Division History, October 1986 - September 1987, Volume I, p. 196, 197; Space Systems Division History, October 1987 - September 1988, Volume I, pp. 306, 309, 315; Space Systems Division History, October 1988 - September 1989, Volume I, p. 342.

21. Space Division History, October 1986 - September 1987, Volume I, pp. 198, 199; Space Systems Division History, October 1988 - September 1989, Volume I, pp. 191, 192.

22. ESMC History, 1 October 1984- 30 September 1986, Volume I, p. 39; ESMC History, 1 October 1987 - 30 September 1988, Volume I, pp. 102, 103, 104; ESMC History, 1 October 1988 - 30 September 1989, Volume I, pp. 161, 162.

23. ESMC History, 1 October 1987 - 30 September 1988, Volume I, pp. 104, 105; ESMC History, 1 October 1988 - 30 September 1989, Volume I, p. 361.

24. ESMC History, 1 October 1987 - 30 September 1988, Volume I, p. 106; ESMC History, 1 October 1988 - 30 September 1989, Volume I, pp. 161, 167, 168, 364.

25. ESMC History, 1 October 1988 - 30 September 1989, Volume I, pp. 163, 166.

26. ESMC History, 1 October 1988 - 30 September 1989, Volume I, p. 166.

27. ESMC History, 1 October 1988 - 30 September 1989, Volume I, pp. 166, 167.

28. ESMC History, 1 October 1989 - 30 September 1990, Volume I, pp. 309; ESMC/45 SPW History, 1 October 1990 - 31 December 1991, Volume I, pp. 155, 156; 45 SPW History, 1 January

- 31 December 1992, Volume I, pp. 227, 229, 230; Range Pretest Briefing, CSR, "NAV II-8," 23 July 1990, pp. 9, 10.

29. ESMC History, 1 October 1988 - 30 September 1989, Volume I, pp. 364, 365, 366; ESMC History, 1 October 1989 - 30 September 1990, Volume I, pp. 311, 312, 313, 314; ESMC/45 SPW History, 1 October 1990 - 31 December 1991, Volume I, pp. 334, 337.

30. ESMC/45 SPW History, 1 October 1990 - 31 December 1991, Volume I, pp. 337, 338, 339; 45 SPW History, 1 January - 31 December 1992, Volume I, pp. 230, 232; Summary, ETR/ROS, "Major Operations FY-92," 4 November 1992, p. 2; Range Pretest Briefing, CSR, "DELTA II LOSAT," 26 January 1990, p. 2.

31. 45 SPW History, 1 January - 31 December 1992, Volume I, pp. 232, 234, 236, 237, 238.

32. Space Systems Division History, October 1988 - September 1989, Volume I, pp. 335, 342, 349.

33. Space Systems Division History, October 1988 - September 1989, Volume I, pp. 342, 346; Space Systems Division History, October 1989 - September 1990, Volume I, pp. 468, 469, 470; Space Systems Division History, October 1990 - September 1991, Volume I, pp. 257, 261; 45 SPW History, 1 January - 31 December 1992, Volume I, pp. 230, 232, 234, 236, 237, 238; Gulf War Air Power Survey, "Space Operations," Vol IV, p. 123.

34. ESMC History, 1 October 1988 - 30 September 1989, Volume I, p. 162; Range Pretest Briefing, Pan Am World Services, Inc. and RCA International Service Corporation, "DELTA-180," August 1986, pp. 1, 2; ESMC History, 1 October 1984 - 30 September 1986, Volume I, pp. 283, 284.

35. ESMC History, 1 October 1987 - 30 September 1988, Volume I, pp. 294, 295.

36. ESMC History, 1 October 1987 - 30 September 1988, Volume I, p. 296; Range Pretest Briefing, Pan Am World Services, Inc. and RCA International Service Corporation, "DELTA-181," January 1988, pp. 1, 3.

37. ESMC History, 1 October 1988 - 30 September 1989, Volume I, pp. 366, 368.

38. Range Pretest Briefing, CSR, "DELTA II LOSAT," 26 January 1990, pp. 1, 3; ESMC History, 1 October 1989 - 30 September 1990, Volume I, pp. 314, 315, 316.

39. ESMC/45 SPW History, 1 October 1990 - 31 December 1991, Volume I, pp. 341, 342.

40. Range Pretest Briefing, Pan Am World Services, Inc. and RCA International Services Corporation, "AC-66, FLTSATCOM F-7," October 1986, pp. 1, 2; Range Pretest Briefing, Pan Am

World Services, Inc. and RCA International Services Corporation, "AC-67, FLTSATCOM F-6," 11 March 1987, pp. 1, 2; Information Summary, NASA/KSC, "FLTSATCOM-G," 4 December 1986; Information Summary, NASA/KSC, "FLTSATCOM-F," 26 March 1987; ESMC History, 1 October 1986 - 30 September 1987, Volume I, pp. 359, 360, 362.

41. ESMC History, 1 October 1986 - 30 September 1987, Volume I, pp. 363, 364, 365, 367, 368.

42. ESMC History, 1 October 1986 - 30 September 1987, Volume I, pp. 368, 369; ESMC History, 1 October 1988 - 30 September 1989, Volume I, pp. 354.

43. ESMC History, 1 October 1988 - 30 September 1989, Volume I, pp. 356, 357, 358.

44. ESMC History, 1 October 1989 - 30 September 1990, Volume I, pp. 108, 109.

45. ESMC History, 1 October 1988 - 30 September 1989, Volume I, pp. 174, 175, 178.

46. ESMC History, 1 October 1988 - 30 September 1989, Volume I, pp. 181, 182; ESMC History, 1 October 1989 - 30 September 1990, Volume I, pp. 112, 113; ESMC/45 SPW History, 1 October 1990 - 31 December 1991, Volume I, pp. 164, 165, 166; 45 SPW History, 1 January - 31 December 1992, Volume I, pp. 86, 87, 222.

47. ESMC/45 SPW History, 1 October 1990 - 31 December 1991, Volume I, p. 331; 45 SPW History, 1 January - 31 December 1992, Volume I, pp. 221, 222.

48. 45 SPW History, 1 January - 31 December 1992, Volume I, pp. 222, 223; Range Operation Briefing, CSR, "ATLAS II/DSCS III Launch, AC-101," January 1992, pp. 2, 3, 4; Memo, Dr Timothy C. Hanley, SMC/HO, 13 October 1993.

49. 45 SPW History, 1 January - 31 December 1992, Volume I, pp. 225, 226, 15.

50. Cleary, *The 6555th*, p. 175; ESMC History, 1 October 1986 - 30 September 1987, Volume I, pp. 146, 147, 149.

51. ESMC History, 1 October 1987 - 30 September 1988, Volume I, pp. 131, 132; ESMC History, 1 October 1988 - 30 September 1989, Volume I, pp. 219, 220, 221, 223.

52. ESMC History, 1 October 1989 - 30 September 1990, Volume I, pp. 144, 147.

53. Range Pretest Briefing, CSR, "STARBIRD Development Launch," December 1990, p. 2; Article, "Starbird rocket takes flight," *Florida Today*, 18 December 1990; ESMC/45 SPW History, 1 October 1990 - 31 December 1991, Volume I, pp. 346, 348, 349.

54. Range Pretest Briefing, CSR, "STARBIRD Development Launch," December 1990, p. 2; Article, "Starbird rocket takes flight," *Florida Today*, 18 December 1990; ESMC/45 SPW History, 1 October 1990 - 31 December 1991, Volume I, p. 349.

55. ESMC/45 SPW History, 1 October 1990 - 31 December 1991, Volume I, pp. 352, 353.

56. ESMC/45 SPW History, 1 October 1990 - 31 December 1991, Volume I, pp. 354, 355.

Future Space Operations

The National Space Transportation and Support Study

On 25 February 1985, President Ronald Reagan signed National Security Decision Directive 261, "Recommendations for Increasing United States Heavy-Lift Space Launch Capacity." The document articulated the need for a national space transportation study, and it directed the Defense Department and NASA to get together to study the development of a second generation space transportation system employing both manned and unmanned launch systems in the post-1995 era. Another directive entitled "National Space Transportation and Support Study (NSTSS)" followed on 14 May 1985, whereupon a joint NASA/Defense Department steering group was established to begin the work in earnest. The steering group was co-chaired by Mr. Jesse W. Moore (NASA's Associate Administrator for Space Flight) and Air Force Under Secretary Edward C. Aldridge, Jr. The principal objective of the study was to identify alternative launch vehicle technologies considered necessary or "prudent" to pursue for the U. S. space program after 1995. The study was also expected to identify potential space mission classes of vehicles, assess support capabilities and discuss "non-vehicle specific technologies" that would help develop the vehicles and support systems needed after 1995.[1]

As its first task, the NSTSS Joint Steering Group was directed to compile a set of potential space mission classes. Its second task involved trade studies, systems analyses and technology assessments to define space transportation options for two sets of space missions-military and civilian. A third task required the identification of technology investment strategies that might "revitalize the nation's launch technology base." The Steering Group's fourth task was to identify technology development programs that could make the entire process pay off at the appropriate time. Though this was a joint effort, the civilian part of the study would be published apart from the Defense Department study to ensure that overlapping needs and differences between military and civilian missions were identified explicitly. On 13 June 1985, a NASA/Defense Department Joint Study Group was set up under the direction of Mr. Paul Holloway and Colonel William F. H. Zersen to implement those civilian and military studies. The Civil Needs Working Group (under the Joint Study Group) published Volume I of the Civil Needs study on 30 October 1985. The Joint Study Group produced a draft of its National Space Transportation and Support Study in May 1986.[2]

In its October 1985 study, NASA noted that the mid to late 1990s offered the nation an opportunity to shift from the Space Shuttle to a low-Earth orbiting space station. The space station was seen as the "cornerstone of an expanding space infrastructure" which could include co-orbiting unmanned platforms as well as polar and geosynchronous unmanned stations. The study also suggested possible landmark missions in the 21st Century, including a manned geosynchronous space station, a habitable lunar base and a Mars sample return. Such missions would require routine inspection, servicing and repair of spacecraft, the assembly of large space systems, the use of Orbital Transfer Vehicles (OTVs) for routine cargo management in space and (most importantly) rapid and effective development of technology to make the entire space infrastructure workable. It must be emphasized that the study presented those ideas on the premise of an expanding space program, *not* the U.S. space program as it was funded when the study was published.[3]

The draft of the National Space Transportation and Support Study went further. Since the level of space operations in the 1995-2010 period was difficult (if not impossible) to predict in 1986, the Joint Study Group's Mission Requirements Panel developed four alternative sets of mission needs which were presented as five combinations of mission models labeled "cases." Case 1 presented a very modest civilian space effort and a constrained military space program. Case 2 was a normal growth program. Case 3 combined the Case 2 scenario with a modest expansion in the civilian sector and a "representative" deployment of SDI kinetic energy weapons. Case 4 combined the Case 2 civilian program with a full-blown SDI program, and Case 5 combined the Case 2 military program with an aggressive civilian space program. Apart from the five case scenarios, the Joint Study Group agreed that, if nothing new was done, the U.S. would enter the late 1990s with: 1) a modest Space Shuttle fleet, 2) TITAN IVs to assure a launch capability for critically important Defense Department missions and 3) a limited launch and spacecraft processing capability. Even under Case 2 conditions, the old space transportation architecture would have a hard time meeting normal growth requirements after the turn of the century. By the year 2005, the first-generation Shuttles would be at the end of their operational life cycles. Several more Shuttle orbiters and many more unmanned vehicles would have to be purchased together with additional support facilities. Despite billions of dollars of additional investment, technological advances would be slow, and the operational savings would be modest.[4]

The Joint Study Group believed there were several alternatives to the existing space transportation architecture. The Unmanned Cargo Vehicle (UCV) and a second-generation Space Shuttle were proposed as twin centerpieces in a modernized space program. Three options were presented in the May 1986 study. Option A included a new partially reusable UCV in the mid-1990s, a partially or fully reusable new upper stage in the late 90s and a Shuttle II vehicle around 2002. Option B included the features of Option A plus a payload return capability featuring payload canisters, a payload glide housing or a fully reusable second stage with a cargo bay. Option C also reflected the characteristics of Option A, but it included modularized elements for the UCV to make deployment of a full-blown SDI network more cost-effective. Elaborating on its investment strategy, the Joint Study Group presented three architectures keyed to options A, B and C. Architecture 1 suggested use of first-generation Space Shuttles through 2002. The Shuttle II vehicle would pick up the manned space mission after 2002, and unmanned payload requirements could be handled by a 150,000-pound payload capacity UCV. Architecture 2 would have the UCV mentioned in Architecture 1, but it would also have a fully recoverable booster and a partially recoverable upper stage. Architecture 3 featured Shuttle operations through 2001, followed by Shuttle II operations beginning in 2002. From 1995 through 2008,

Architecture 3 called for a 150,000-pound payload capacity UCV with a payload deorbit capability as well as a launch capability. To provide greater cost-effectiveness, the UCV would be equipped with a fully reusable first stage booster in 2008. Though the annual investment in any one of the three architectures could run as high as $7 billion, the Joint Study Group believed that all three showed lower life cycle costs than the existing fleet of Shuttles and unmanned launch vehicles. The Group deemed Architecture 2 the most cost-effective alternative. Architecture 1 was considered the least cost-effective choice.[5]

Future Space Operations

The Space Transportation Architecture Study and Advanced Launch System (ALS) Studies

In addition to the long-term goals of the National Space Transportation and Support Study, the government wanted to examine payload requirements and launch options projected over the middle term (e.g., 1985-1995). Toward that end, a jointly funded NASA, Air Force and SDIO Space Transportation Architecture Study (STAS) was offered in June 1985, and four contracts worth approximately $6,000,000 apiece were awarded to Boeing, Rockwell International, General Dynamics and Martin Marietta on 6 September 1985. Not surprisingly, the result of this effort was very modest: a special report in December 1986 discussed SDI requirements, and a final Interim Progress Report made basic recommendations concerning space architecture in the 1990s. In the short term, the launch community could only hope to sort out cost reduction opportunities for the TITAN IV, DELTA II and Space Shuttle. Any major reductions in operating costs awaited a more thorough examination of space support systems, followed by proposals for an advanced launch system.[6]

In July 1987, one-year concept definition contracts were issued to seven contractors to prepare the groundwork for the Advanced Launch System (ALS). As a result of that effort, a better picture of future launch systems emerged: the new ALS vehicle would mostly likely be based on a hydrogen-fueled core with a varying number of strap-on solid rocket or liquid engine boosters. By taking this "tinker toy" (modular) approach to the vehicle, payloads ranging from 10,000 to 200,000 pounds could be lifted into space much more economically. Borrowing a page from Soviet and Western European rocket design, the new vehicles would be simpler and heavier, thus avoiding the expensive "high performance" characteristics of American boosters in the past. Three basic booster concepts dominated the contractors' proposals. The least costly vehicle would employ a hydrogen core and from six to twelve solid rocket boosters. The core stages would provide all thrust vector control, and the solids would be designed with composite, monolithic cases and fixed rocket nozzles. A more costly vehicle employed a liquid core and from one to six strap-on liquid rocket engines. All core and strap-on stages would have common components, tankage and structure, and all engines would be ignited on the launch stand to verify proper operation before the vehicle was released (i.e., no in-flight engine starts would be required). The most expensive and most advanced ALS alternative was a winged, fully reusable booster. Owing to technological uncertainty, the winged booster was not as likely a candidate for the ALS as the other two booster concepts.[7]

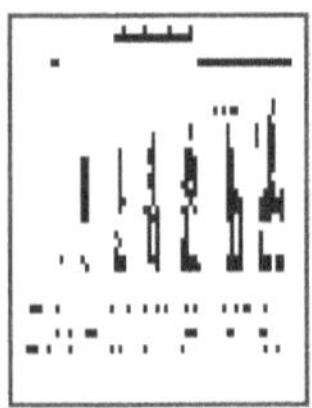

Figure 136: Reference ALS Family of Vehicles (Conceptual)

Ground support facilities were addressed in the ALS studies as well. The old launch pads with their fixed umbilical towers would be replaced with austere launch pads devoid of service towers. To avoid tying up the launch pad for long periods of time, most launch processing would occur away from the pad, and payload/launch vehicle mating would be accomplished in a vertical integration facility. Only when the vehicle was ready to launch would it be moved out to the pad, fueled from base mounted utilities and launched. In the contractors' opinion, the key to streamlining the launch process lay in integrating and testing the launch vehicle away from the pad. In the event a vehicle broke down before launch, it could be quickly removed from the pad without costly teardown, transport, reassembly and retest procedures.[8]

Figure 137: ALS Common Core Stage

The Defense Acquisition Board was briefed on the results of the ALS concept definition studies in September 1988. The Board embraced cost reduction as the primary focus for the ALS, and it supported the concept of a family of launch vehicles based on standardized modules. On 4 November 1988, Defense Secretary Frank Carlucci signed an Acquisition Decision Memorandum (ADM) approving the initial concept. This action paved the way for further ALS efforts. In December 1988, three two-year Phase II ALS contracts were awarded to Boeing, General Dynamics and a Martin Marietta/McDonnell Douglas contracting team for the development of ALS designs and demonstrations of ALS technology. Initially, the U.S. Government hoped to find out if the ALS could be expanded to accommodate ten 100,000-pound and twenty 150,000-pound payload class launches per year by 2005. Nevertheless, by early March 1989, a major change in the ALS Systems Requirement Document introduced two different baseline scenarios for the ALS: 1) a smaller vehicle with a launch rate of six to ten vehicles per year around 2000, and 2) a larger booster with a launch rate of 20 or more vehicles by 2009. The Space Systems Division Program Manager for ALS instructed the contractors to design their ALS facilities with payloads up to 220,000 pounds in mind. Furthermore, the facilities were to be sited to allow growth to a 300,000-pound payload class system in the more distant future.[9]

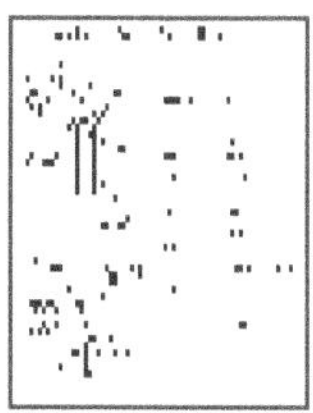

Figure 138: ALS Solid Rocket Motor Description

Political and budgetary realities were already at work modifying prospects for the ALS. In March 1989, the new SDIO Director, Lt. General George L. Monahan, Jr. told the House Armed Services Subcommittee on Research and Development that his office "saw little urgency to build an ALS" before the early decades of the 21st Century. In April, the *Air Force Times* reported a "sentiment in Congress" for a scaled-down ALS using improved versions of present-generation boosters. In the spring and summer of 1989, it seemed less and less likely that the Bush Administration would push a full-blown Strategic Defense Initiative program through Congress or that a future administration would be able to fund such a system. In October 1989, funding cutbacks effectively killed support for the SDIO's Zenith Star (directed energy) experimental mission, and the shift in emphasis to lightweight kinetic energy weapons (e.g., Brilliant Pebbles and Brilliant Eyes) reduced SDIO payload requirements dramatically. Since much of the demand for large ALS boosters depended on a vigorous SDI program as well as a low-Earth orbiting space station, full-scale development of the ALS family of vehicles would have to wait until the Air Force and NASA had substantive mission requirements for the big boosters. In a joint message from the Secretary of the Air Force and Air Force Headquarters on 7 December 1989, the ALS Program Office was directed to terminate Phase II design efforts "as soon as practical" and suspend any new obligations pending redirection of the ALS into a technology program. In early February 1990, the Space Systems Division Commander praised the ALS program as a technical and managerial success. Though the ALS family of boosters would not be developed, all three ALS contractors were directed to transfer ALS technology to the existing fleet of unmanned boosters. The ATLAS, TITAN and IUS programs all stood to benefit from automated ground support systems, off-line processing and "paperless" management techniques encouraged under Phase II. According to NASA's calculations, some of the greatest reductions in launch costs would come from the development of low-cost engines (modestly funded through 1998) and streamlined processing procedures.[10]

The net effect of ALS was to prepare the DELTA II, ATLAS II and TITAN IV programs for further refinements through the 1990s. (The dream of a new generation of launch vehicles was not dead, but, as a practical matter, it would have to wait for another time.) In the meantime, the ALS Program Manager, Colonel John R. Wormington, succeeded Colonel Roy D. Bridges, Jr. as ESMC Commander on 27 January 1990. One of Colonel Wormington's first actions as Commander was to have his plans people begin a study of launch, range and user requirements at the Cape through 2005. Though the effort was led by Plans (ESMC/XR), the study was broken down into functional areas with team leaders from various ESMC agencies to oversee investigations into each area. Following the kick-off meeting on 8 February 1990, five interactive panels began looking at launch processing facilities, range modernization, communications, safety and weather. The interactive panels were backed up with six support panels to address range infrastructure, launch commercialization, manpower, budget, logistics and security. A twelfth panel was created to document the ESMC baseline anticipated by 2005.[11]

Future Space Operations

The ESMC 2005 Study and Prospects for the Current Generation of Unmanned Space Launch Vehicles

According to the ESMC 2005 study plan (dated 6 March 1990), the written product would consist of three volumes: the management overview (Volume I), the narrative summary (Volume II), and a review of the planning processes and conclusions associated with each of the functional areas of study (Volume III). It must be emphasized that the effort was a study, not a plan in the formal sense. It was offered as a vision of the future and a proposal of what might be possible by the year 2005. As a local initiative, it did not enjoy national visibility, and its impact on Air Force decision-makers was debatable. Nevertheless, ESMC 2005 provided another look into the future of space operations at the Cape, and it deserves at least a brief review. For our purposes, highlights from Volume III will suffice to cover the study.[12]

Figure 139: Structure and Feed System Commonality

According to "ESMC 2005," a future land use plan was developed to support the concentration of all launch pads along Heavy Launch Row (i.e., from Complex 46 at the eastern tip of the Cape northward to Complex 41). Under the plan, most of the land north of the Skid Strip and west of Heavy Launch Row would be reserved for vehicle processing and launch support operations. The area south of the Skid Strip would be used for payload processing and any vehicle processing overflow that could not be accommodated north to the Skid Strip. The pattern of development would make the best use of restricted real estate on the Cape, and it would provide the best protection for people and support facilities. In conjunction with the land use plan, the Cape's infrastructure would be expanded to provide new power distribution, roads, water supply and sewage treatment facilities for the new launch vehicle complexes and support facilities arrayed north and south of the Skid Strip. Since most of the existing infrastructure had been funded in piecemeal fashion over the previous four decades, the Cape's new infrastructure would be a major departure from the past. The Infrastructure Support Panel speculated that AFSPACECOM might underwrite the effort with new sources of capital investment. In any event, adequate support for the Cape's workload in 2005 would depend heavily on a modern infrastructure of roads and utilities.[13]

For forty years, the Cape's payload and launch processing facilities had been funded and constructed for specific launch vehicle and spacecraft programs. This customized approach to space missions caused needless duplication of facilities and ground equipment, and it spilled over into launch pad operations where extensive testing could tie up launch pads for weeks at a time. In retrospect, the whole approach was very inefficient and cumbersome. In the ESMC 2005 view, payloads and launch vehicles could be processed in automated "common use" facilities. While assembly and integration facilities might vary for classes of launch vehicles (e.g., small, medium and large), they could all share automated technologies that would remove all major integration activities from the pad. To break the grip of program-specific funding, a central authority would have to be set up to solicit funds to build "generic" complexes (e.g., a new consolidated launch operations control complex, a consolidated payload processing/encapsulation complex and a logistics and launch management center). Launch pads could be standardized for small, medium and large launch vehicles, and vehicle assembly complexes could be developed along those class lines.[14]

In line with the recommendations of the Infrastructure Support Panel, the Launch Processing Facilities Panel recommended Complex 17 be deactivated at the conclusion of the DELTA II program. That action would free up real estate at the south end of the Cape for industrial and administrative buildings. Thereafter, modern medium launch vehicles could lift off from any one of four austere generic launch complexes (e.g., complexes 12, 14, 16 and 36). Three other deactivated complexes (11, 13 and 15) could be reserved for future small launch vehicle operations. Unlike the medium launch pads, each of the small launch pads would get a new umbilical tower to support propellant tanking operations, pressurization and power requirements. Most of the small vehicle checkout and launch equipment would be set up at common use launch control facilities behind earth revetments near the pads. There was also some possibility that the Navy might develop Complex 46 as a small space launch complex in the future.[15]

The Launch Processing Facilities Panel also recommended that complexes 34 and 37 be reserved for Heavy Launch Vehicles (HLVs) for "an indefinite period." Another class of vehicles-Large Launch Vehicles (LLVs)- might be created to bridge the gap between the "old" TITAN IVs and the new modular HLVs. The LLVs could incorporate ALS technology and possibly lift payloads as heavy as 80,000 pounds into low-Earth orbit. If that strategy was followed, complexes 40 and 41 might be modified into LLV complexes around the turn of the century. Later, the Air Force could develop its class of HLVs to boost 100,000-pound to 250,000-pound payloads into orbit from Complex 34 and/or Complex 37.[16]

Figure 140: Map depicting possible Heavy Launch Vehicle Tow Route from the Skid Strip to Pads 34 and 37 (e.g., ALS Pads A and B)

The ESMC 2005 study was not merely a conceptual "wish list." It presented the evolution, costs and potential savings of new launch processing facilities in considerable detail. For TITAN IV launch operations, control was shifting from the Vertical Integration Building to the Launch Operations Control Center when ESMC 2005 began. The study suggested spending $110 million to consolidate DELTA II, ATLAS II and TITAN IV operations in a new Consolidated Launch Operations Control Center (CLOCC) near the new Range Operations Control Center (ROCC) at the south end of the Cape. Such an investment might save $90 million on ATLAS II and DELTA II operations. Another $100 million might be saved by the CLOCC when new medium, large and heavy launch vehicle programs were introduced around the turn of the century. Along the same lines, off-pad integration and storage of fully assembled TITAN IV vehicles could require a new $60 million Final Assembly Building (FAB) and a $16 million vehicle transporter system, but the cost of those initiatives would be more than offset by the savings in downtime due to current vehicle assembly bottlenecks. The introduction of laser-initiated ordnance on TITAN IVs could allow simultaneous on-pad operations at complexes 40 and 41, thereby saving additional processing time.[17]

According to the ESMC 2005 scenario, approximately $2.4 billion might be invested on medium, large and heavy launch vehicle assembly and integration facilities over a seven-year period. Another $100,000,000 would be needed to convert complexes 40 and 41 into research and development facilities after the heavy launch vehicles came on line. By 2005, the study's contributors imagined heavy launch vehicle pads might exist on or near complexes 34 and 37. The new pads would cost about $630 million in 1990 dollars. Modernized medium and large launch vehicle programs could require $310,000,000 more in engineering and support facilities. With regard to payload processing facilities, $250,000,000 might be needed for TITAN IV "off-pad" processing by the turn of the century. Another $45,000,000 might be spent on transportation, storage and fairing assembly requirements for the new off-line TITAN payloads. Large and heavy launch vehicle payload processing might cost another $619,000,000, but the new medium launch vehicle program could benefit from that development and buy its own payload encapsulation capabilities for about $30,000,000.[18]

While many other topics were addressed in the ESMC 2005 study, the preceding paragraphs provide a summary of the proposals likely to have the greatest impact on future space launch operations at the Cape-provided the overall scenario of new large and heavy launch vehicles is adopted eventually. Given the disintegration of the Soviet Union and the increasingly chilly fiscal climate in America in the 1990s, it is by no means certain that any of the study's recommendations will be adopted. More immediate problems challenge U.S. military space operations. As of this writing, the recent flight failure of a TITAN IV at Vandenberg has delayed TITAN/CENTAUR operations at the Cape, and the Solid Rocket

Motor Upgrade (SRMU) has experienced more than its share of development problems in recent years. On the positive side, the military space program has displayed amazing resilience, most notably after the Challenger disaster in 1986. It is fair to say that the TITAN IV program has encountered lengthy, but not insurmountable, delays. If the past is any guide, the TITAN IV will be a feature of the Cape's military space operations for many years to come.[19]

On the other hand, if the circumstances governing military space operations change sufficiently, the nation may be faced with some hard choices. If, for the sake of argument, the TITAN IV/CENTAUR can not provide the reliability the Defense Department needs for its most critical payloads, some military payloads may be manifested as Space Shuttle missions or transferred to a heavy lifting version of the ARIANE, perhaps the ARIANE 5. Another vehicle may be introduced, or the payloads themselves might be downsized to meet changing world conditions. All of those alternatives are within the realm of possibility, though there may be little (if any) direct evidence to bear any of them out. Barring extraterrestrial influences, unmanned launch vehicles with liquid rocket engines and solid rocket motors will be the mainstay of military space operations at the Cape for at least another generation and probably two. A new manned vehicle may eventually replace the old reliable boosters and the Space Shuttle in two generations, but there are many technological unknowns associated with that approach, and there will be questions concerning a new manned system's reliability, if the Space Shuttle is any guide. In the short and middle term, we keep coming back to the current generation of ATLAS II, DELTA II and TITAN IV launch vehicles. If their manufacturers can continue to launch them safely, successfully and in a reasonably timely fashion, medium and heavy unmanned boosters should have a bright future at the Cape. If not, other alternatives will have to be pursued.

Chapter Four Footnotes

The steering group
The other members of the group included: Dr. William R. Lucas (Director, Marshall Space Flight Center) Dr. Norman Terrell (NASA Associate Administrator for Policy), Dr. Raymond S. Colladay (NASA Associate Administrator for Aeronautics and Space Technology), Lt. General Bernard P. Randolph, (USAF Deputy Chief of Staff, Research, Development and Acquisition), Lt. General James A. Abrahamson (Director, SDIO), and Dr. Larry L. Woodruff (Office of the Under Secretary for Defense Research and Engineering-Strategic and Theater Nuclear Forces). Two executive secretaries were also appointed to the steering group as non-voting members. They were Ivan Bekey (for NASA) and Dr. Thomas P. Rona (for the Defense Department).

characteristics of American boosters
American rocket design had been based on minimum weight criteria designed to get the most out of every pound of rocket fuel. While that principle kept the vehicle's weight down, it required the use of expensive lightweight materials and high-pressure components. The Western European and Soviet philosophy suggested vehicle weight advantages could be traded for lower cost: a heavier simpler booster could accomplish the same job for less money overall.

Phase II ALS contracts
The Martin Marietta/McDonnell Douglas Phase II contract was valued at $97,900,000. The Boeing and General Dynamics contracts were worth $82,900,000 and $82,700,000 respectively.

funding cutbacks
The FY 1989 ALS budget was cut 38 percent after ALS launch requirements were reduced to reflect a more modest ballistic missile defense network of kinetic energy weapons. The Advanced Launch System's initial funding profile-predicated on the deployment of larger and heavier directed energy weapons-was no longer valid.

Complex 17
Range Safety drew Impact Limit Lines (ILL) around Complex 17 to protect workers from the potential dangers of DELTA launch failures at the Cape. In effect, the lines prohibited the development of conventional structures a considerable distance from pads 17A and 17B. Once the launch danger was removed, inhabited "soft" structures could be constructed closer to the pads.

development problems
Two Class A mishaps in September 1990 and April 1991 hindered the SRMU's development at the Astronautics Laboratory at Edwards Air Force Base, California. Though the mishaps had no direct

impact on the Cape, they translated into program delays that inevitably retarded the introduction of the SRMU on TITAN IVs on the Eastern Range. Happily, the SRMU's first full duration firing was completed successfully on 12 June 1992 at the Phillips Laboratory (formerly the Astronautics Laboratory). The first qualification static test firing on 15 October 1992 was also successful. Hercules President Richard Schwartz was quoted in an official Air Force news release as feeling "highly optimistic about the preliminary results." Additional static firing tests was completed successfully on 12 September 1993.

Chapter Four Endnotes

1. NSTSS, "Volume I, Civil Needs, Final Report," 30 October 1985, 1-1, 1-2, 1-3; NSTSS, "National Space Transportation Strategy, 1995-2010," (Draft), May 1986, Foreword.

2. ESMC History, 1 October 1986 - 30 September 1987, Volume I, pp. 115, 116, 117; NSTSS, "Civil Needs," Volume I, 30 October 1985, pp. 3-4, 3-5, 3-6, 3-7; NSTSS, "National Space Transportation Study, 1995-2010," (Draft) May 1986, pp. 1, 2, 3, 4, 5, 6.

3. *Ibid.*

4. NSTSS, "National Space Transportation Strategy, 1995-2010," (Draft) May 1986, pp. 12, 13, 14, 22, 24, 26, 27.

5. NSTSS, "National Space Transportation Strategy, 1995-2010," (Draft) May 1986, pp. 39, 40, 41, 54, 55, 57.

6. ESMC History, 1 October 1986 - 30 September 1987, Volume I, pp. 128, 129, 133, 135; Space Systems Division History, October 1988 - September 1989, Volume I, p. 139.

7. ESMC History, 1 October 1987 - 30 September 1988, Volume I, p. 110; Space Systems Division History, October 1988 - September 1989, Volume I, pp. 140, 143, 144.

8. Space Systems Division History, October 1988 - September 1989, Volume I, pp. 143, 144.

9. Space Systems Division History, October 1988 - September 1989, Volume I, pp. 142, 144; ESMC History, 1 October 1988 - 30 September 1989, Volume I, pp. 118; ESMC History, 1 October 1989 - 30 September 1990, Volume I, pp. 356, 357.

10. ESMC History, 1 October 1989 - 30 September 1990, Volume I, pp. 148, 359, 361, 362.

11. ESMC History, 1 October 1989 - 30 September 1990, Volume I, pp. 67, 369, 370.

12. Letter, Lt. Colonel Ernest L. Lockwood, ESMC/XR, to ESMC/CC et al, "ESMC 2005 Requirements Model," 13 March 1990; Summary, ESMC/XR, "Study Plan for ESMC 2005," 6 March 1990.

13. Study, ESMC/XR, " ESMC 2005 Functional Area Requirements Technology Data," 24 July 1990, Volume III, p. 7-9.

14. Study, ESMC/XR, "ESMC 2005 Functional Area Requirements Technology Data," 24 July 1990, Volume III, pp. 1-3, 2-3, 2-4, 2-5.

15. Study, ESMC/XR, "ESMC 2005 Functional Area Requirements Technology Data," 24 July 1990, Volume III, p. 2-11.

16. Study, ESMC/XR, "ESMC 2005 Functional Area Requirements Technology Data," 24 July 1990, Volume III, p. 2-12; Interview, M. C. Cleary with Mr. Ed Herrburger, ESMC/XRX, 17 September 1990.

17. Study, ESMC/XR, "ESMC 2005 Functional Area Requirements Technology Data," 24 July 1990, Volume III, Launch Processing Facilities Summary (following p. 2-12).

18. *Ibid.*

19. 45 SPW History, 1 January - 31 December 1992, Volume I, pp. 218, 219. 45 SPW History, 1 January - 31 December 1993, pp 191,192.

134

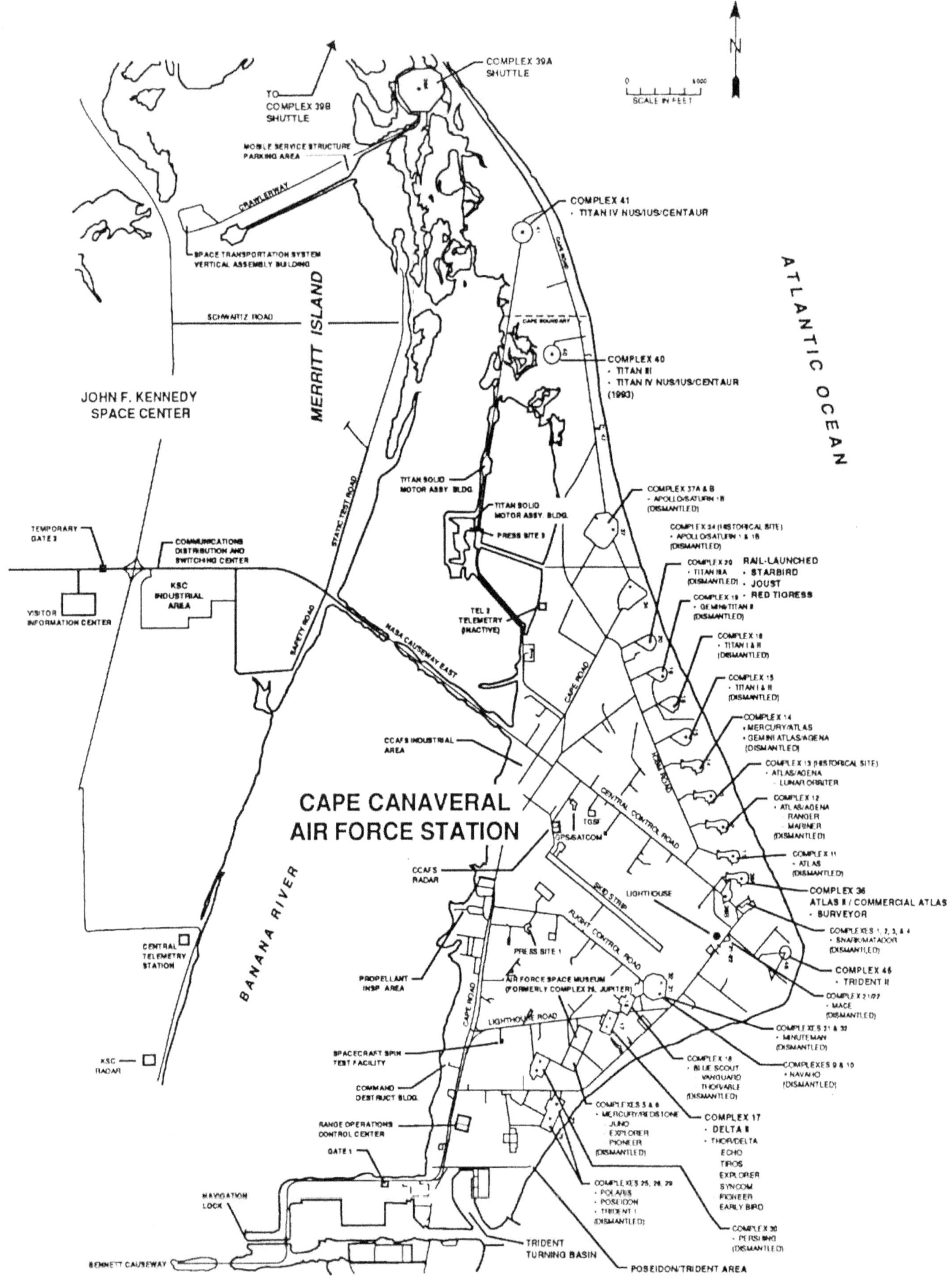
N
SCALE IN FEET
COMPLEX 39A
SHUTTLE
TO COMPLEX 39B
SHUTTLE
MOBILE SERVICE STRUCTURE
PARKING AREA
CRAWLERWAY
SPACE TRANSPORTATION SYSTEM
VERTICAL ASSEMBLY BUILDING
COMPLEX 41
· TITAN IV NUS/IUS/CENTAUR
ATLANTIC OCEAN
MERRITT ISLAND
SCHWARTZ ROAD
CAPE BOUNDARY
COMPLEX 40
· TITAN III
· TITAN IV NUS/IUS/CENTAUR
(1993)
JOHN F. KENNEDY
SPACE CENTER
STATIC TEST ROAD
TITAN SOLID
MOTOR ASSY. BLDG.
TITAN SOLID
MOTOR ASSY. BLDG.
PRESS SITE 3
COMPLEX 37A & B
· APOLLO/SATURN 1B
(DISMANTLED)
COMPLEX 34 (HISTORICAL SITE)
· APOLLO/SATURN 1 & 1B
(DISMANTLED)
TEMPORARY
GATE 3
COMMUNICATIONS
DISTRIBUTION AND
SWITCHING CENTER
KSC
INDUSTRIAL
AREA
VISITOR
INFORMATION CENTER
COMPLEX 20
· TITAN IIIA
(DISMANTLED)
RAIL-LAUNCHED
· STARBIRD
· JOUST
· RED TIGRESS
COMPLEX 19
· GEMINI/TITAN II
(DISMANTLED)
TEL II
TELEMETRY
(INACTIVE)
SAFETY ROAD
NASA CAUSEWAY EAST
CAPE ROAD
COMPLEX 16
· TITAN I & II
(DISMANTLED)
COMPLEX 15
· TITAN I & II
(DISMANTLED)
COMPLEX 14
· MERCURY/ATLAS
· GEMINI ATLAS/AGENA
(DISMANTLED)
CCAFS INDUSTRIAL
AREA
COMPLEX 13 (HISTORICAL SITE)
· ATLAS/AGENA
· LUNAR ORBITER
CAPE CANAVERAL
AIR FORCE STATION
CENTRAL CONTROL ROAD
TGSF
GPS/SATCOM
COMPLEX 12
· ATLAS/AGENA
RANGER
MARINER
(DISMANTLED)
COMPLEX 11
· ATLAS
(DISMANTLED)
CCAFS
RADAR
BANANA RIVER
SKID STRIP
LIGHTHOUSE
COMPLEX 36
ATLAS II / COMMERCIAL ATLAS
· SURVEYOR
COMPLEXES 1, 2, 3, & 4
· SNARK/MATADOR
(DISMANTLED)
CENTRAL
TELEMETRY
STATION
PRESS SITE 1
FLIGHT CONTROL ROAD
COMPLEX 46
· TRIDENT II
PROPELLANT
INSP AREA
AIR FORCE SPACE MUSEUM
(FORMERLY COMPLEX 26, JUPITER)
COMPLEX 21/22
· MACE
(DISMANTLED)
LIGHTHOUSE ROAD
COMPLEXES 31 & 32
· MINUTEMAN
(DISMANTLED)
KSC
RADAR
SPACECRAFT SPIN
TEST FACILITY
COMPLEX 18
· BLUE SCOUT
VANGUARD
THOR/ABLE
(DISMANTLED)
COMPLEXES 9 & 10
· NAVAHO
(DISMANTLED)
COMMAND
DESTRUCT BLDG.
COMPLEXES 5 & 6
· MERCURY/REDSTONE
JUNO
EXPLORER
PIONEER
(DISMANTLED)
COMPLEX 17
· DELTA II
· THOR/DELTA
ECHO
TIROS
EXPLORER
SYNCOM
PIONEER
EARLY BIRD
RANGE OPERATIONS
CONTROL CENTER
GATE 1
COMPLEXES 25, 26, 29
· POLARIS
· POSEIDON
· TRIDENT 1
(DISMANTLED)
NAVIGATION
LOCK
COMPLEX 30
· PERSHING
(DISMANTLED)
TRIDENT
TURNING BASIN
BENNETT CAUSEWAY
POSEIDON/TRIDENT AREA

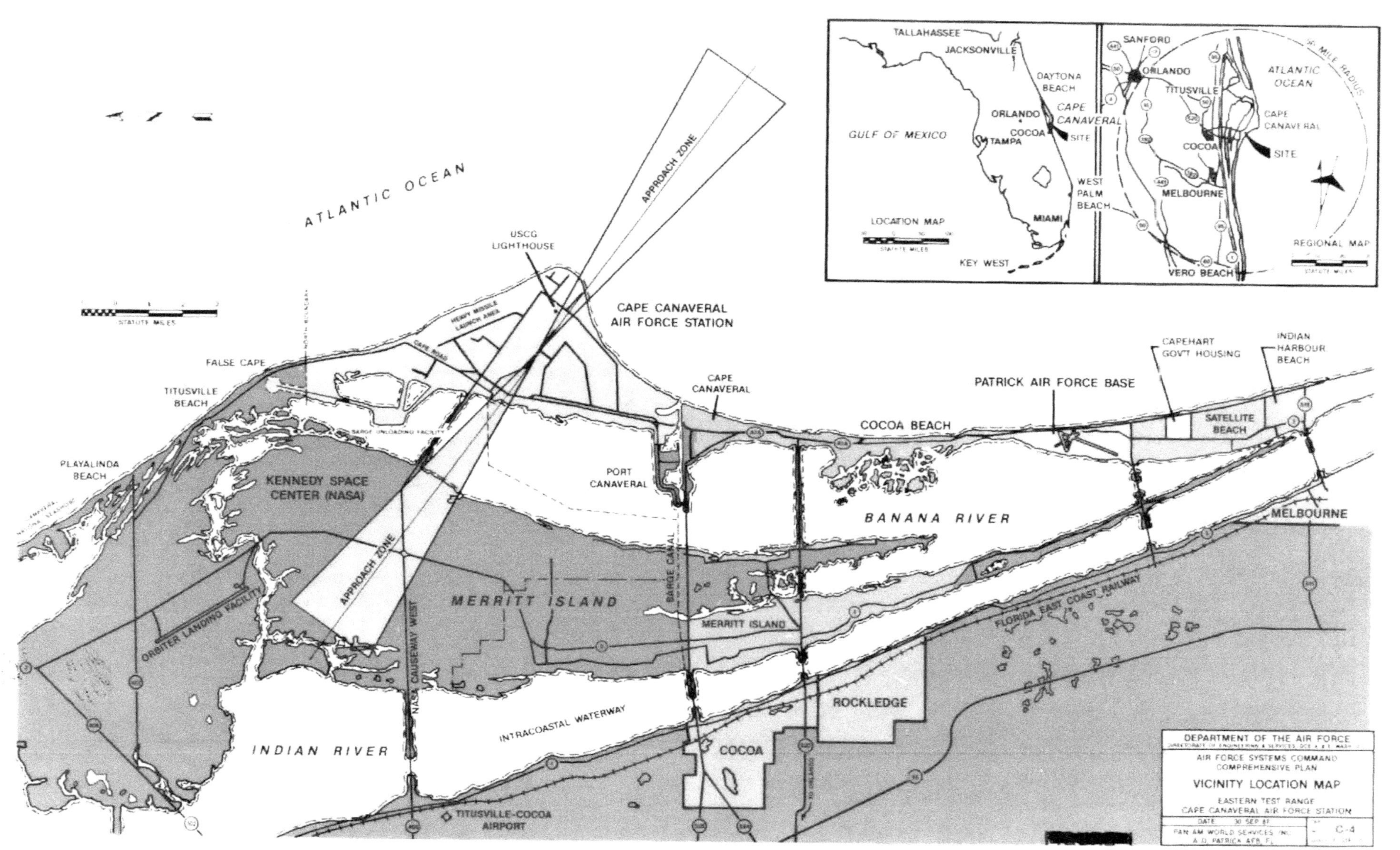
ATLANTIC OCEAN
APPROACH ZONE
USCG LIGHTHOUSE
CAPE CANAVERAL AIR FORCE STATION
HEAVY MISSILE LAUNCH AREA
CAPE ROAD
FALSE CAPE
TITUSVILLE BEACH
BARGE UNLOADING FACILITY
PLAYALINDA BEACH
KENNEDY SPACE CENTER (NASA)
PORT CANAVERAL
CAPE CANAVERAL
COCOA BEACH
PATRICK AIR FORCE BASE
CAPEHART GOV'T HOUSING
INDIAN HARBOUR BEACH
SATELLITE BEACH
BANANA RIVER
MELBOURNE
MERRITT ISLAND
BARGE CANAL
ORBITER LANDING FACILITY
NASA CAUSEWAY WEST
INTRACOASTAL WATERWAY
INDIAN RIVER
FLORIDA EAST COAST RAILWAY
ROCKLEDGE
COCOA
TO ORLANDO
TITUSVILLE-COCOA AIRPORT
STATUTE MILES
TALLAHASSEE
JACKSONVILLE
DAYTONA BEACH
ORLANDO
COCOA
CAPE CANAVERAL
SITE
GULF OF MEXICO
TAMPA
WEST PALM BEACH
MIAMI
KEY WEST
LOCATION MAP
SANFORD
TITUSVILLE
ATLANTIC OCEAN
50 MILE RADIUS
MELBOURNE
VERO BEACH
REGIONAL MAP
DEPARTMENT OF THE AIR FORCE
AIR FORCE SYSTEMS COMMAND
COMPREHENSIVE PLAN
VICINITY LOCATION MAP
EASTERN TEST RANGE
CAPE CANAVERAL AIR FORCE STATION
DATE 30 SEP 87
PAN AM WORLD SERVICES INC
A.D. PATRICK AFB, FL
C-4

FIRE

22-3

USAF

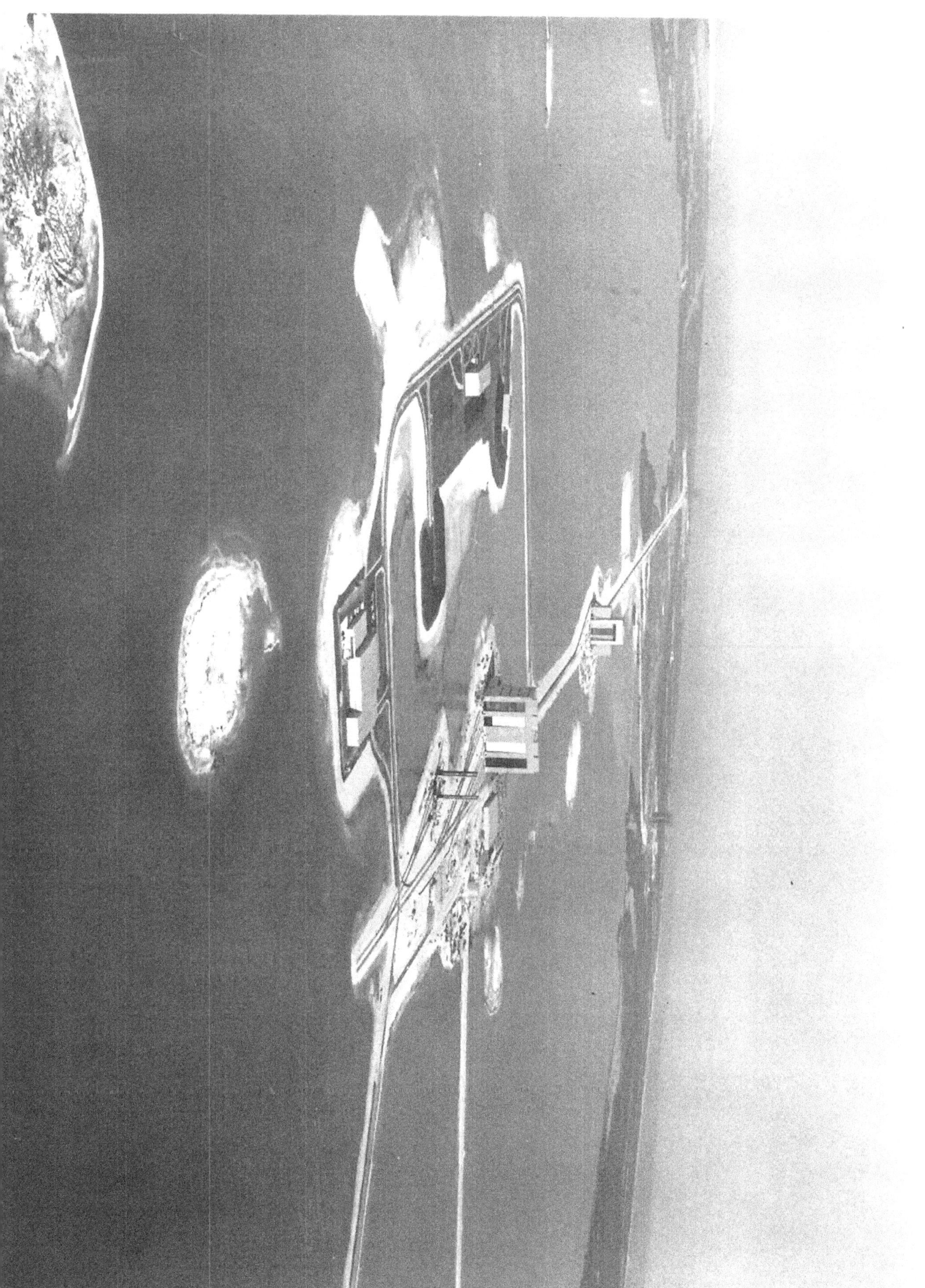

1st SPACE LAUNCH SQUADRON

HRC 90-187

2ND SPACE LAUNCH SQUADRON

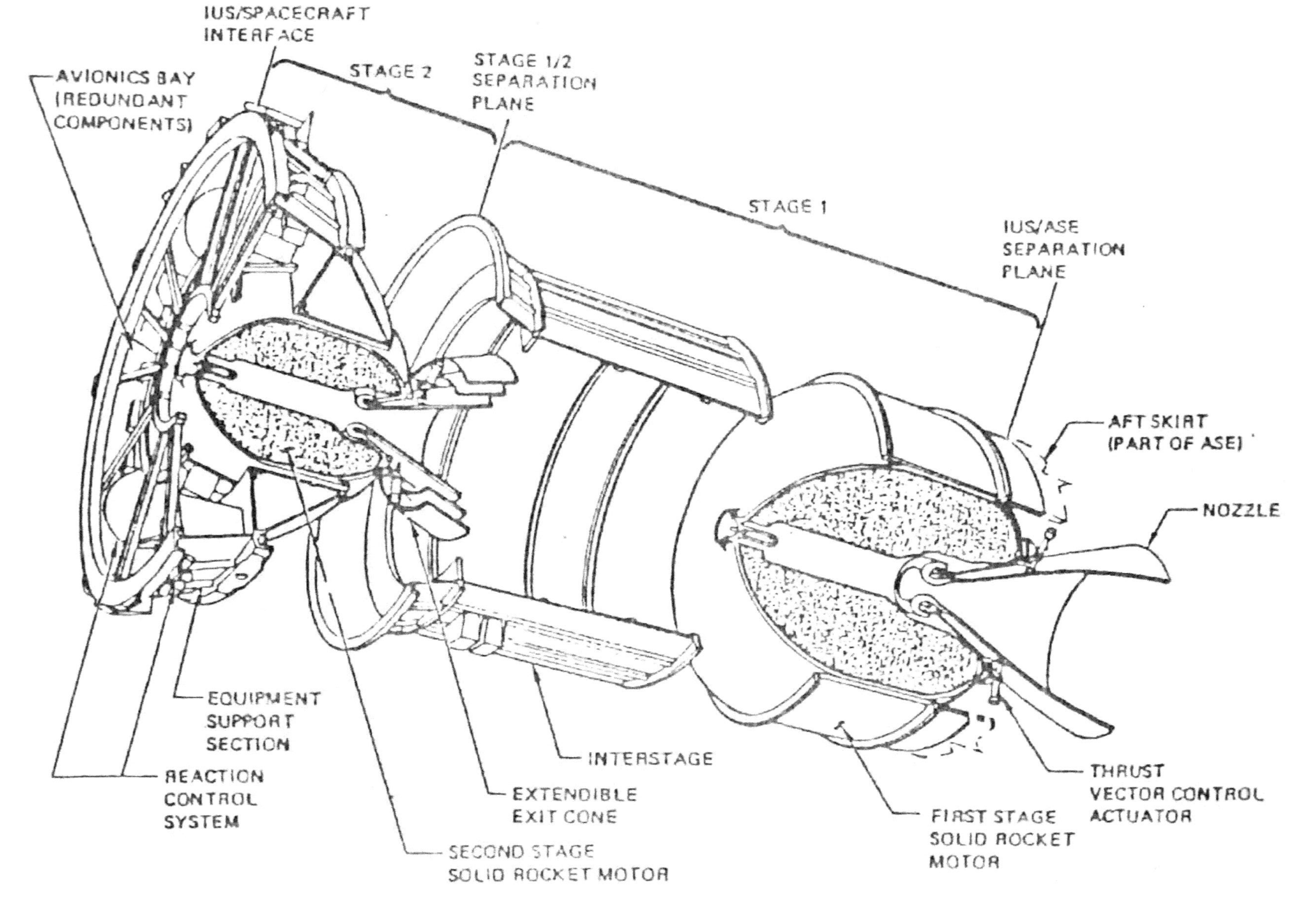
IUS/SPACECRAFT INTERFACE
AVIONICS BAY (REDUNDANT COMPONENTS)
STAGE 2
STAGE 1/2 SEPARATION PLANE
STAGE 1
IUS/ASE SEPARATION PLANE
AFT SKIRT (PART OF ASE)
NOZZLE
EQUIPMENT SUPPORT SECTION
REACTION CONTROL SYSTEM
INTERSTAGE
EXTENDIBLE EXIT CONE
SECOND STAGE SOLID ROCKET MOTOR
FIRST STAGE SOLID ROCKET MOTOR
THRUST VECTOR CONTROL ACTUATOR

MOBILE SERVICE TO
ACCESS TOWER
SHUTTLE ASSEMBLY BUILDING
USAF
PAYLOAD CHANGEOUT ROOM
MERGENCY EGRESS ANDING ZONE
ISS SHELTER
FVV STACK
EXHAUST DUCTS
LIQUID OXYGEN TANK
PAYLOAD PREPARATION ROOM
WATER TREATMENT HOLDING TANK

UNITED STATES

AIR FORCE

U. S. AIR FORCE
COMPLEX - 13
6555 TH ASTG
SPACE LAUNCH VEHICLES SYSTEMS DIVISION

UNITED
47
STATES

UNITED STATES
66
ATLAS
CENTAUR

U.S. AIR
TITAN III

United States

USAF
MDAC
DELTA

U.S.
U.S.

SPENCE

45TH OPERATIONS GROUP

LAUNCH VEHICLE DATA CENTER

ENTRANCE

Console	Equipment	Occupant
01	2, 3, 4, 8	
02 ·	2, 3, 5, 6, 7, 8	
03	2, 3, 4, 8	SPO MISSION MNGR
04	2, 3, 4, 8	AEROSPACE
05	1, 3, 5, 7, 8, 10, 11	AEROSPACE
06 ·	2, 3, 4, 6, 8	MDSSC
07	2, 4, 8, 9	
08 ·	2, 3, 5, 6, 7, 8	
09	2, 3, 4, 8	MDSSC
10	2, 3, 4, 7, 8	RESERVED
11	2, 4, 8	AEROSPACE
12.	2, 3, 5, 6, 7, 8	MDSSC
13.	2, 3, 4, 8	MDSSC
14	2, 4, 8, 9	ROCKETDYNE
15.	1, 3, 5, 7, 8, 10, 11	AEROJET
16	2, 3, 4, 8	RESERVED
17	2, 3, 4, 8	ASST BOOSTER COUNTDOWN CONTROLLER
18	2, 3, 4, 8	AEROSPACE
19	2, 3, 4, 8	
20.	2, 5, 8	
21	2, 3, 4, 8	MDSSC
· 22	2, 3, 4, 7, 8	MDSSC
· 23	2, 3, 5, 6, 7, 8, 12	MDSSC
· 24	2, 3, 4, 8, 12	AEROSPACE
25	2, 4, 8, 9	AEROSPACE
26	1, 4, 5, 7, 8, 10, 11	SSD/CLZ
· 27	2, 4, 8	MDSSC
28	2, 4, 7, 8, 12	MDSSC - CCAFS
· 29	2, 4, 5, 7, 8, 12	MDSSC
30	2, 4, 8, 9	
31	1, 10, 11	FACILITY ANOMALY CHIEF
32	5, 8, 10, 11	BOOSTER COUNTDOWN CONTROLLER
33	1, 7, 10, 11	ANOMALY TEAM CHIEF
34	1, 7, 10, 11	MDSSC
35	5, 8, 10, 11	MDSSC RESERVED
36	1, 5, 11	MDSSC
37	1, 10	
38	1, 10, 11	
39	5, 7, 8	MDSSC
40	1, 10, 11	MDSSC
41	4, 8	
42	1, 10, 11	MDSSC - CCAFS
43	1, 10, 11	AFQA
44.	1, 7, 8	AFQA
45	4, 7, 8	
46.	1, 10, 11	MDSSC - CCAFS
47	1, 10, 11	MDSSC - CCAFS
48		MDSSC WX

TELEPHONES

EXIT

1 = VCSS
2 = BLUE TOPS
3 = 17" MONITOR
4 = 14" MONITOR
5 = 9" MONITOR
6 = RDS TERM
7 = RTSS TERM
8 = VIDEO SW
9 = HARD COPY
10 = GRN PHN
11 = COM CONT PNL
12 = CCC PNL
· = CARDS

MISSION DIRECTOR CENTER

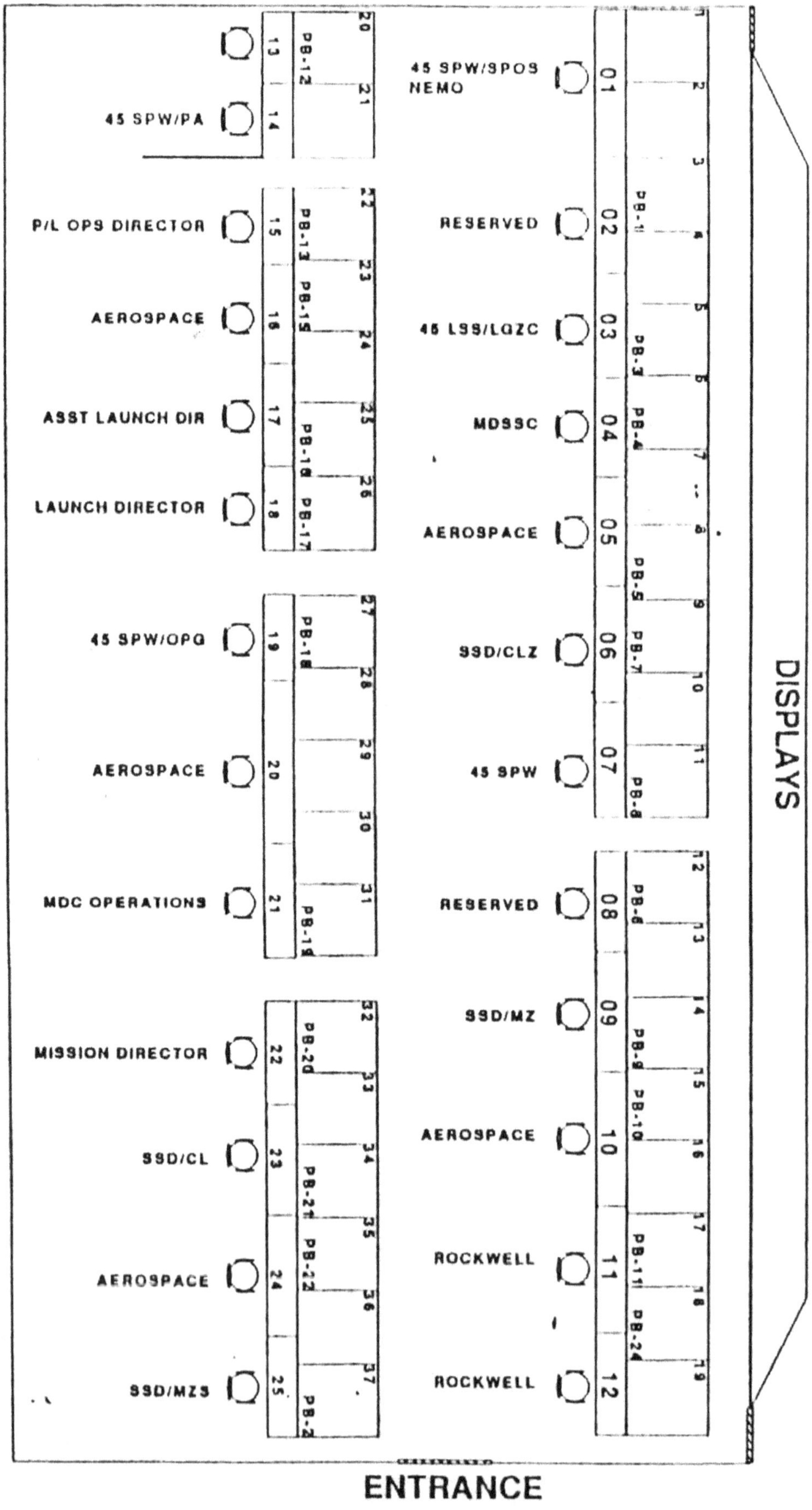

45 SPW/PA

P/L OPS DIRECTOR

AEROSPACE

ASST LAUNCH DIR

LAUNCH DIRECTOR

45 SPW/OPG

AEROSPACE

MDC OPERATIONS

MISSION DIRECTOR

SSD/CL

AEROSPACE

SSD/MZS

45 SPW/SPOS NEMO

RESERVED

45 LSS/LGZC

MDSSC

AEROSPACE

SSD/CLZ

45 SPW

RESERVED

SSD/MZ

AEROSPACE

ROCKWELL

ROCKWELL

DISPLAYS

ENTRANCE

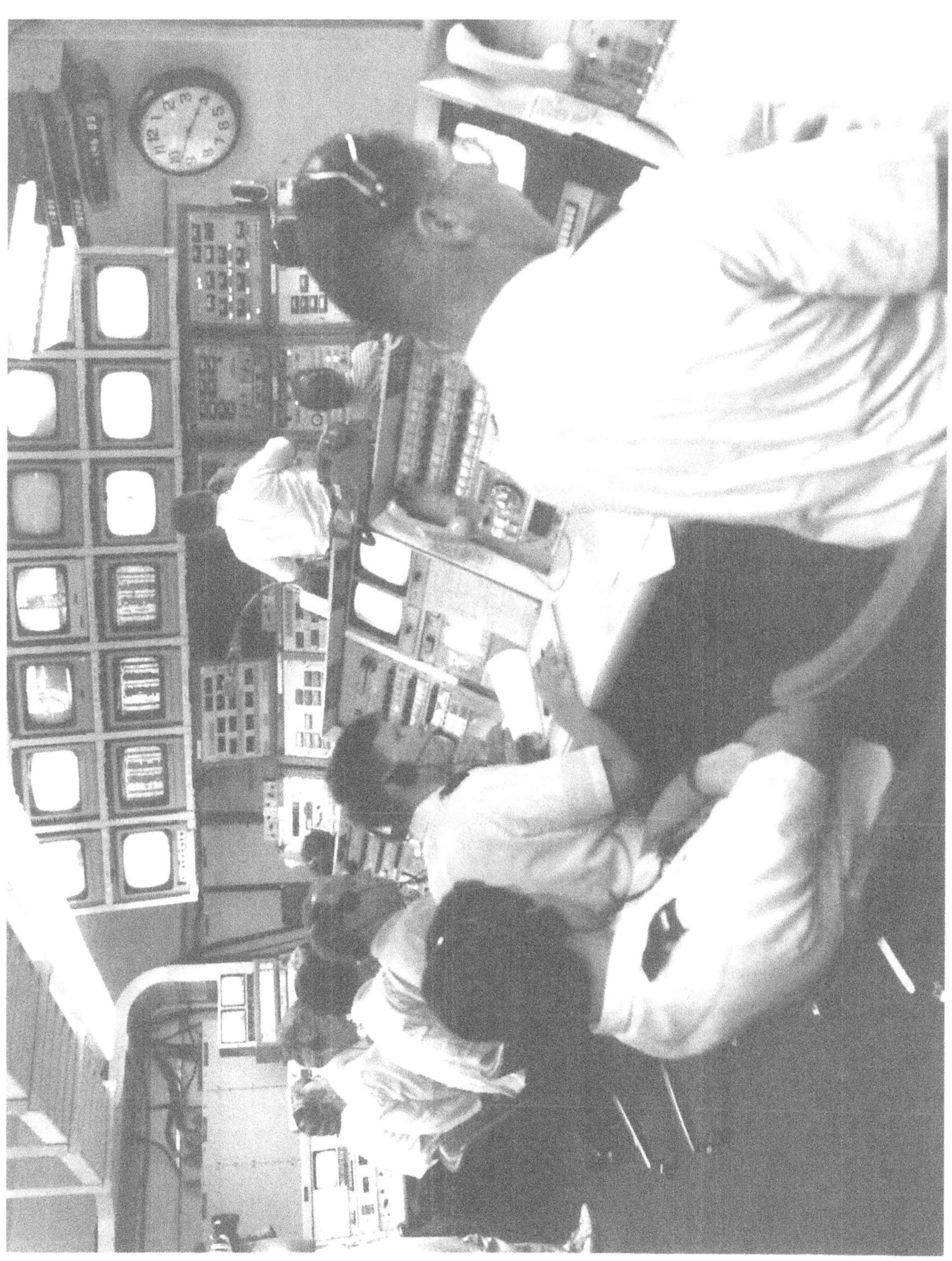

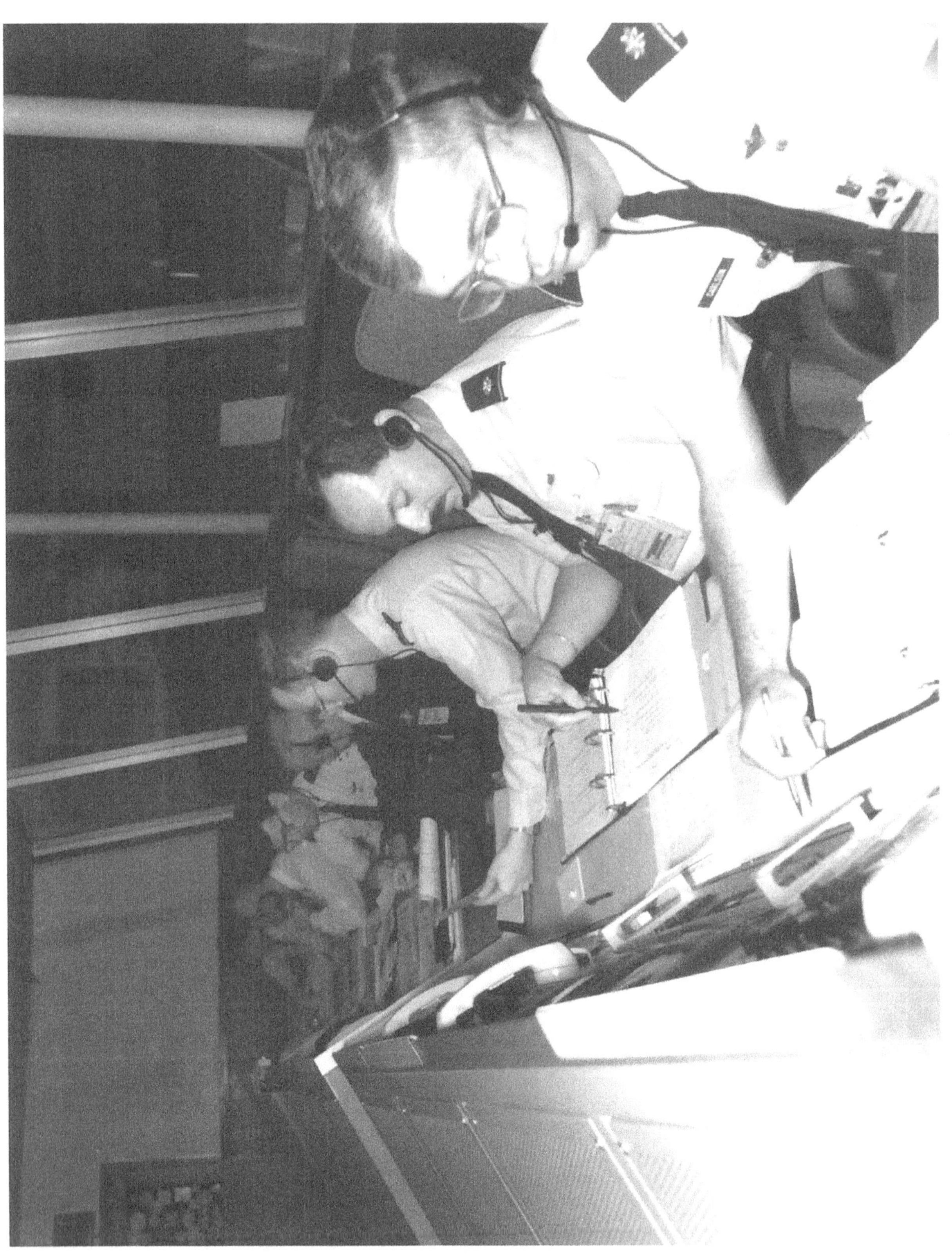

BOEING
IUS

WOODWARD
U.S.

INTEGRATE - TRANSFER - LAUNCH (ITL) FACILITY SHOWING TITAN IIIC LIFT-OFF

AIR FORCE
COMPLEX

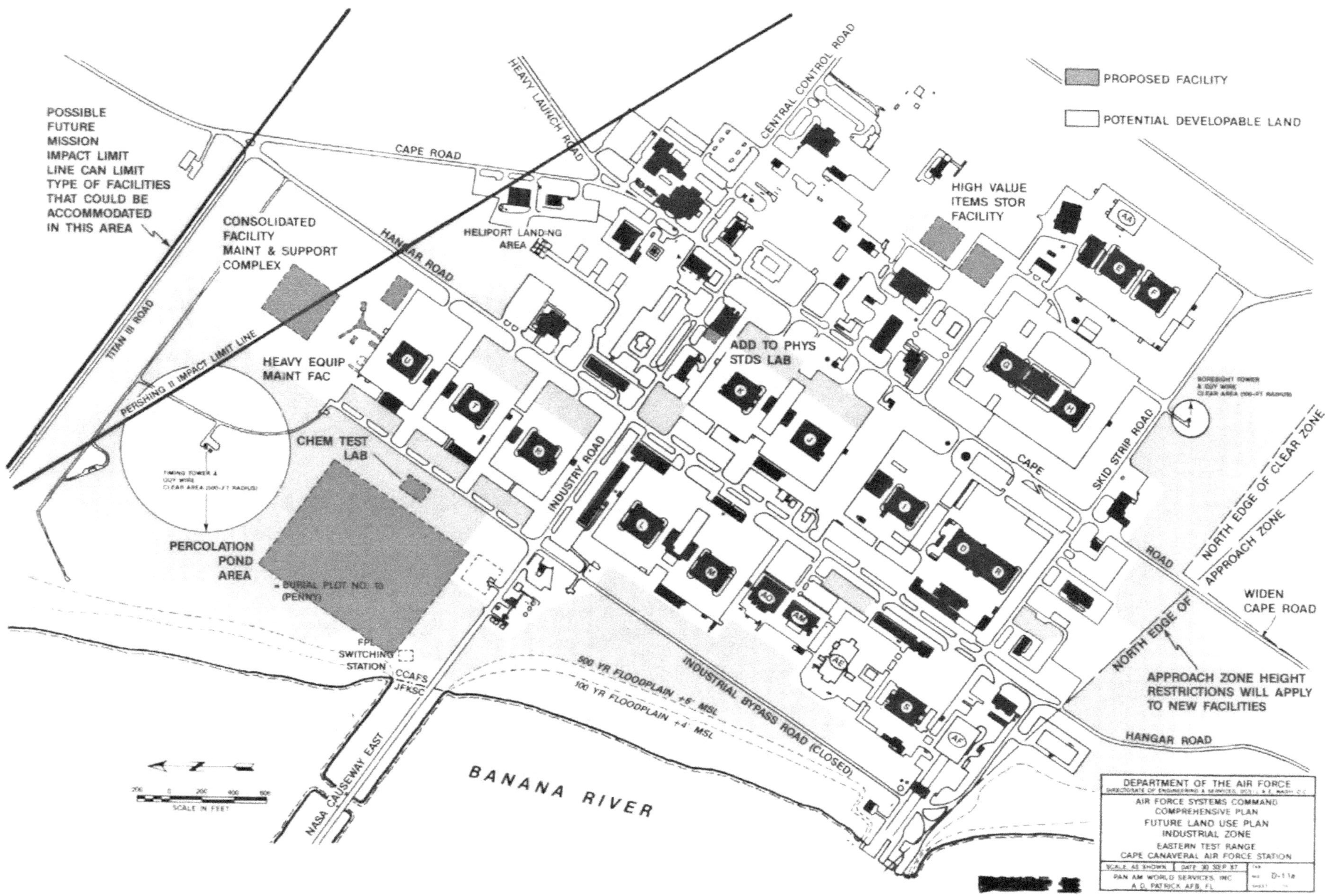
PROPOSED FACILITY
POTENTIAL DEVELOPABLE LAND
POSSIBLE FUTURE MISSION IMPACT LIMIT LINE CAN LIMIT TYPE OF FACILITIES THAT COULD BE ACCOMMODATED IN THIS AREA
CAPE ROAD
HEAVY LAUNCH ROAD
CENTRAL CONTROL ROAD
HIGH VALUE ITEMS STOR FACILITY
CONSOLIDATED FACILITY MAINT & SUPPORT COMPLEX
HELIPORT LANDING AREA
HANGAR ROAD
TITAN III ROAD
PERSHING II IMPACT LIMIT LINE
HEAVY EQUIP MAINT FAC
ADD TO PHYS STDS LAB
CHEM TEST LAB
INDUSTRY ROAD
CAPE
SKID STRIP ROAD
ROAD
NORTH EDGE OF CLEAR ZONE
APPROACH ZONE
WIDEN CAPE ROAD
PERCOLATION POND AREA
BURIAL PLOT NO. 10 (PENNY)
NORTH EDGE OF
APPROACH ZONE HEIGHT RESTRICTIONS WILL APPLY TO NEW FACILITIES
FPL SWITCHING STATION
CCAFS
JFKSC
500 YR FLOODPLAIN +5' MSL
100 YR FLOODPLAIN +4' MSL
INDUSTRIAL BYPASS ROAD (CLOSED)
HANGAR ROAD
NASA CAUSEWAY EAST
BANANA RIVER
SCALE IN FEET
DEPARTMENT OF THE AIR FORCE
AIR FORCE SYSTEMS COMMAND
COMPREHENSIVE PLAN
FUTURE LAND USE PLAN
INDUSTRIAL ZONE
EASTERN TEST RANGE
CAPE CANAVERAL AIR FORCE STATION
SCALE: AS SHOWN
DATE: 30 SEP 87
PAN AM WORLD SERVICES, INC.
A.O. PATRICK AFB, FL
D-11a

U.S.

U.S.
YAGER

U.S. AIR FORCE
6555 TH
AEROSPACE
TEST GROUP

MARTIN

AIR FORCE
COMPLEX 40

AIR FORCE

US AIR FORCE

U.S.
KUHLMAN

U.S.
MARTINELLI

Carlisle

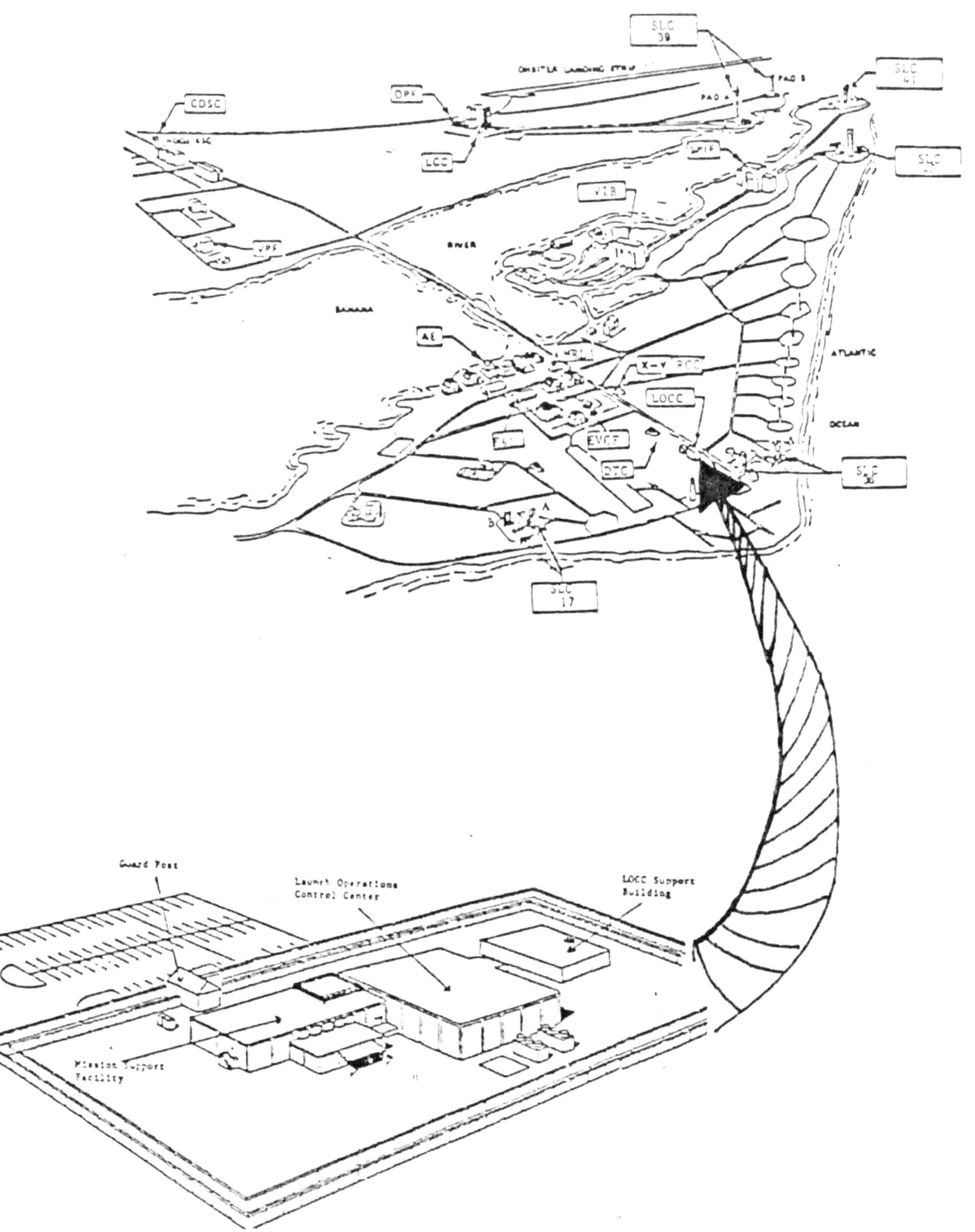

SLC 39
SLC 41
SLC 40
SLC 36
SLC 17
CDSC
OPF
LCC
VPF
VIB
AE
MRL
X-Y RCC
LOCC
EVCF
DTC
PAD A
PAD B
RIVER
BANANA
ATLANTIC
OCEAN
A
B
Guard Post
Launch Operations Control Center
LOCC Support Building
Mission Support Facility

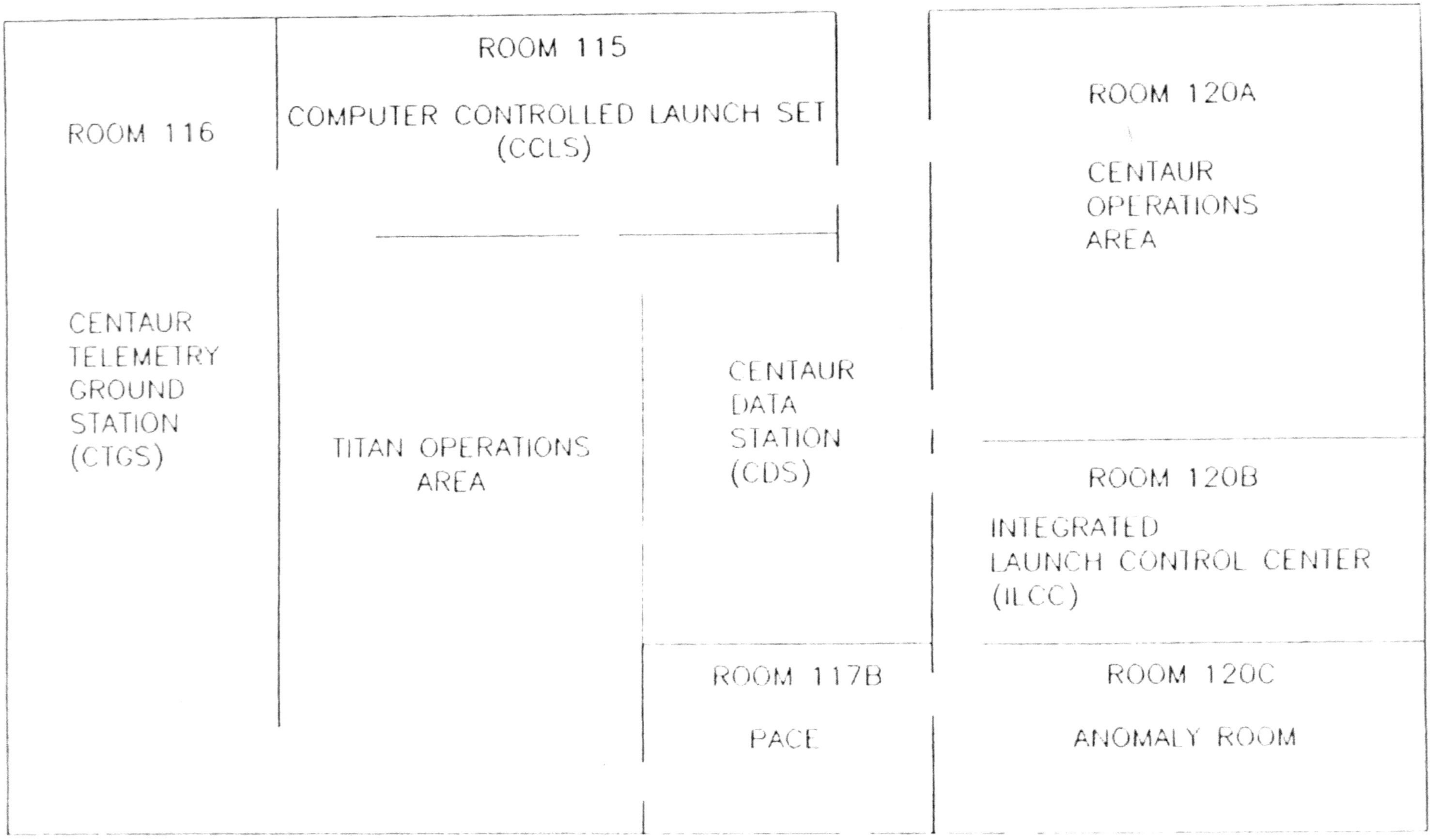

ROOM 116

CENTAUR
TELEMETRY
GROUND
STATION
(CTGS)

ROOM 115

COMPUTER CONTROLLED LAUNCH SET
(CCLS)

TITAN OPERATIONS
AREA

CENTAUR
DATA
STATION
(CDS)

ROOM 117B

PACE

ROOM 120A

CENTAUR
OPERATIONS
AREA

ROOM 120B

INTEGRATED
LAUNCH CONTROL CENTER
(ILCC)

ROOM 120C

ANOMALY ROOM

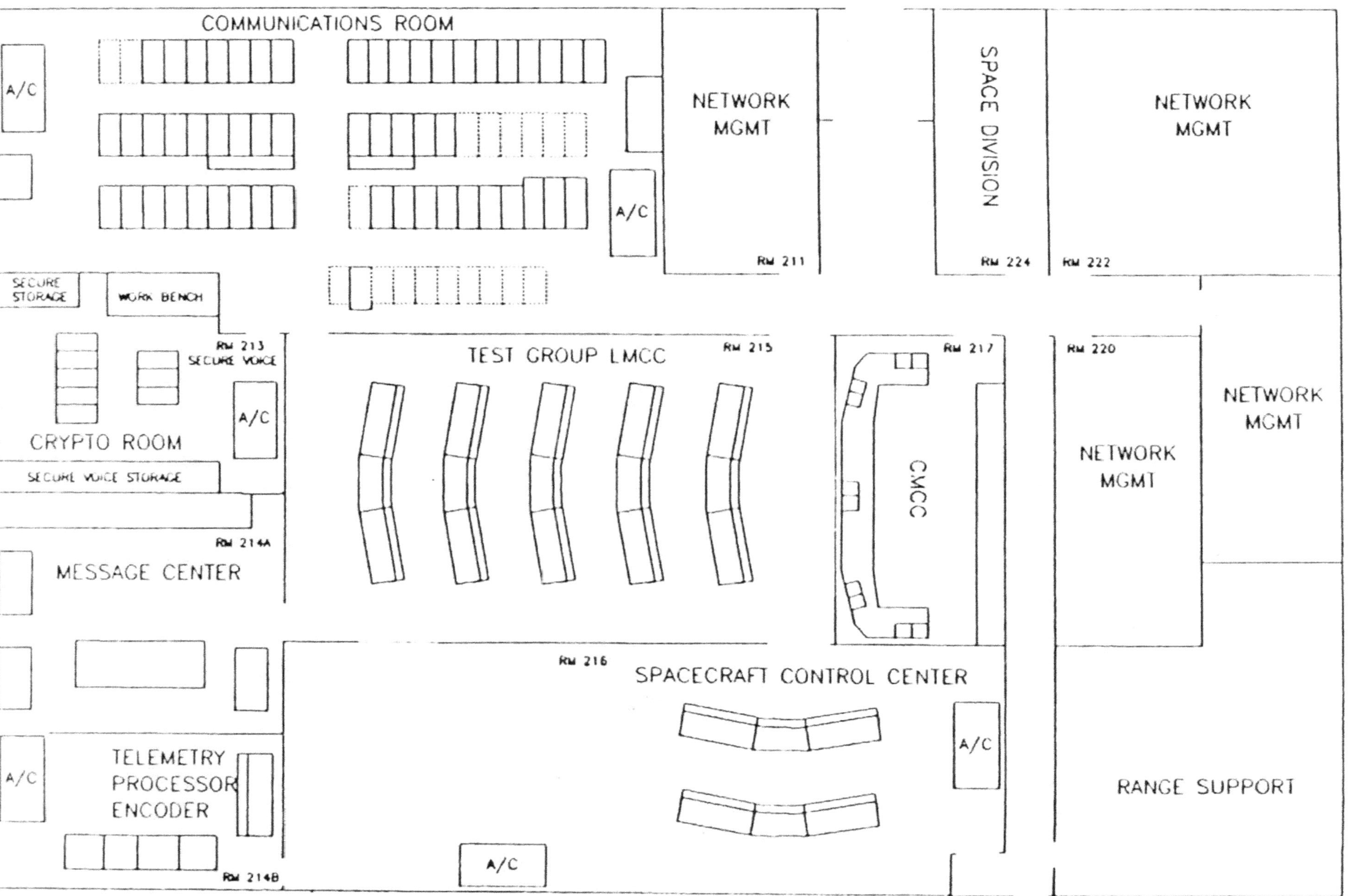
COMMUNICATIONS ROOM
A/C
NETWORK MGMT
RM 211
SPACE DIVISION
RM 224
NETWORK MGMT
RM 222
SECURE STORAGE
WORK BENCH
RM 213
SECURE VOICE
A/C
CRYPTO ROOM
SECURE VOICE STORAGE
TEST GROUP LMCC
RM 215
RM 217
CMCC
RM 220
NETWORK MGMT
NETWORK MGMT
RM 214A
MESSAGE CENTER
RM 216
SPACECRAFT CONTROL CENTER
A/C
A/C
TELEMETRY
PROCESSOR
ENCODER
RANGE SUPPORT
RM 214B
A/C
A/C

US AIR FORCE

MILITARY AIRLIFT COMMAN

U.S.
U.S.
BOURNE

U S AIR FORCE
COMPLEX 41

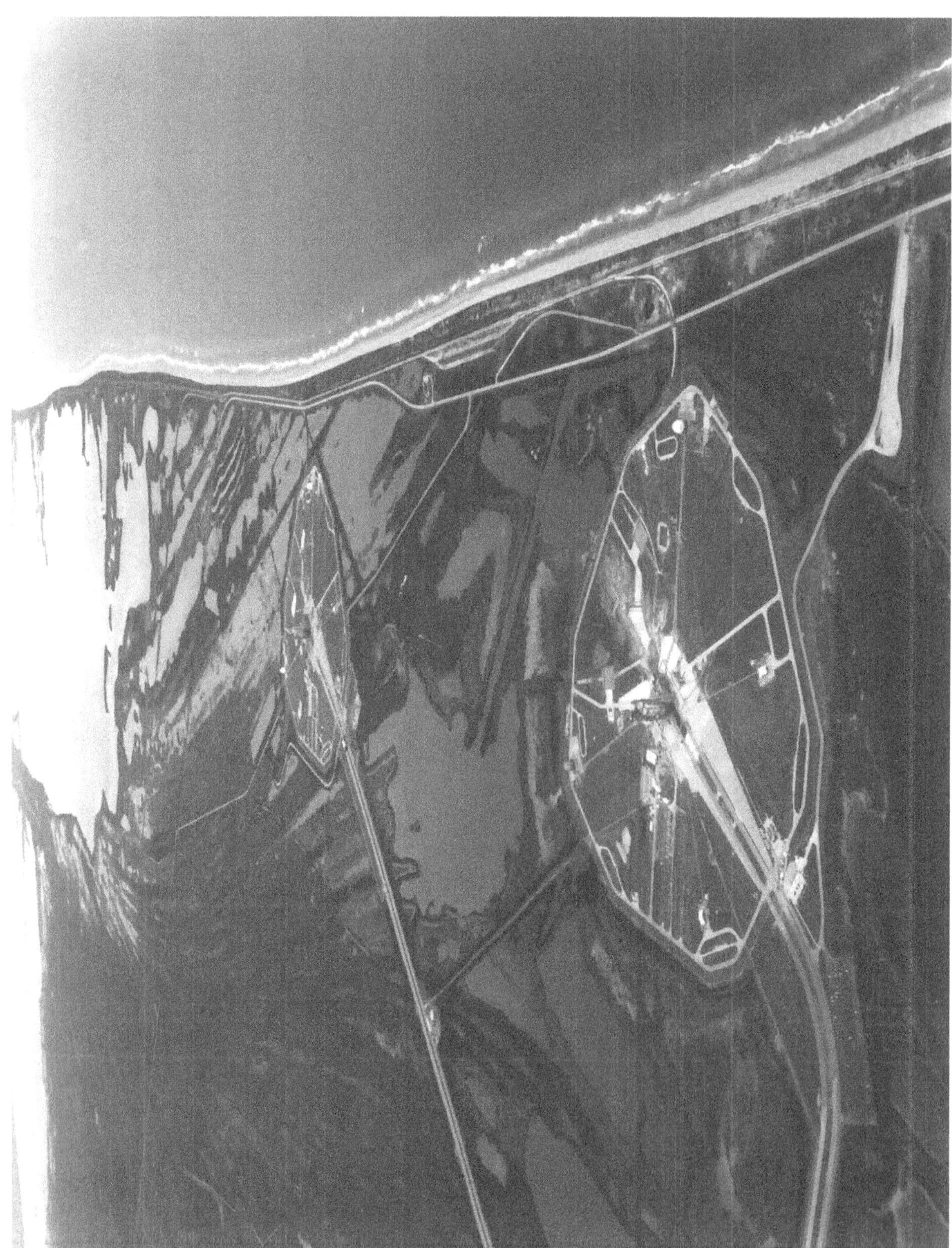

USA

NATO

NATO

UNITED
44
STATES

NITROGEN

NASA
DELTA
NATO

S

UNITED STATES

USAF
DELTA

NASA
DELTA

NASA
DELTA

USAF
MDAC
DELTA
II
B

ATLAS
CENTAUR

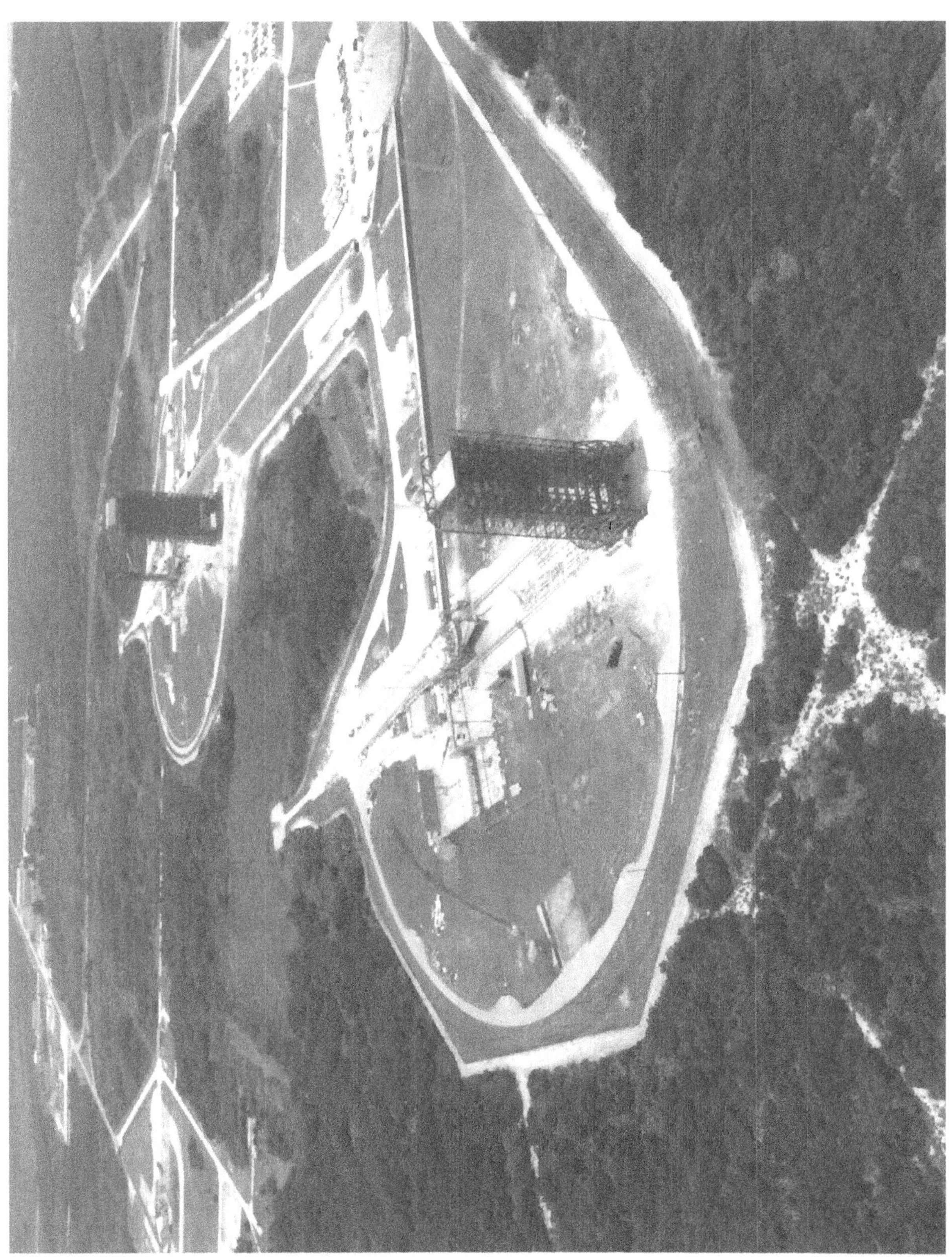

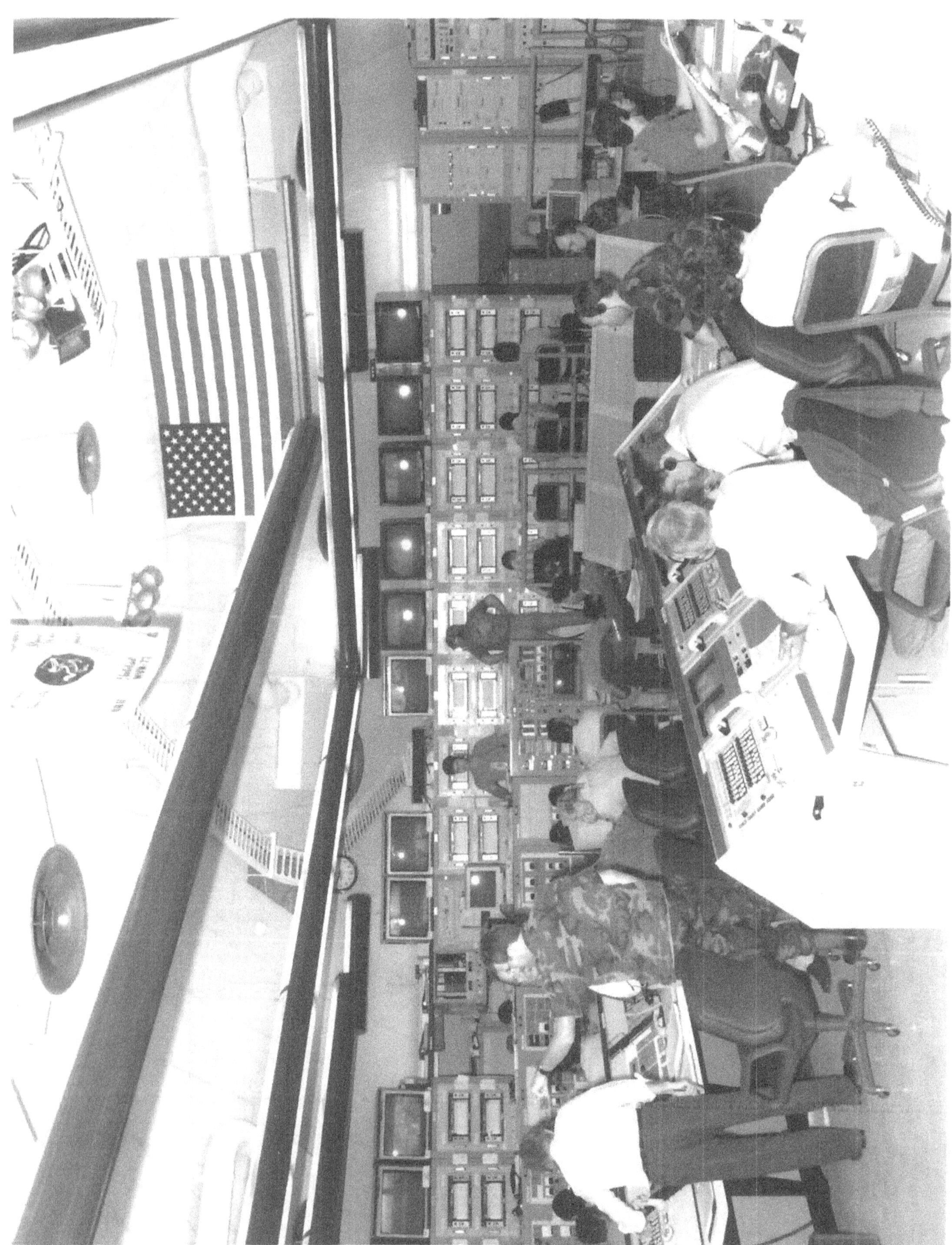

GENERAL
DYNAMICS
GENERAL
103

3D SPACE LAUNCH SQUADRON

TARBIRD

LOADED

THE ALS FAMILY

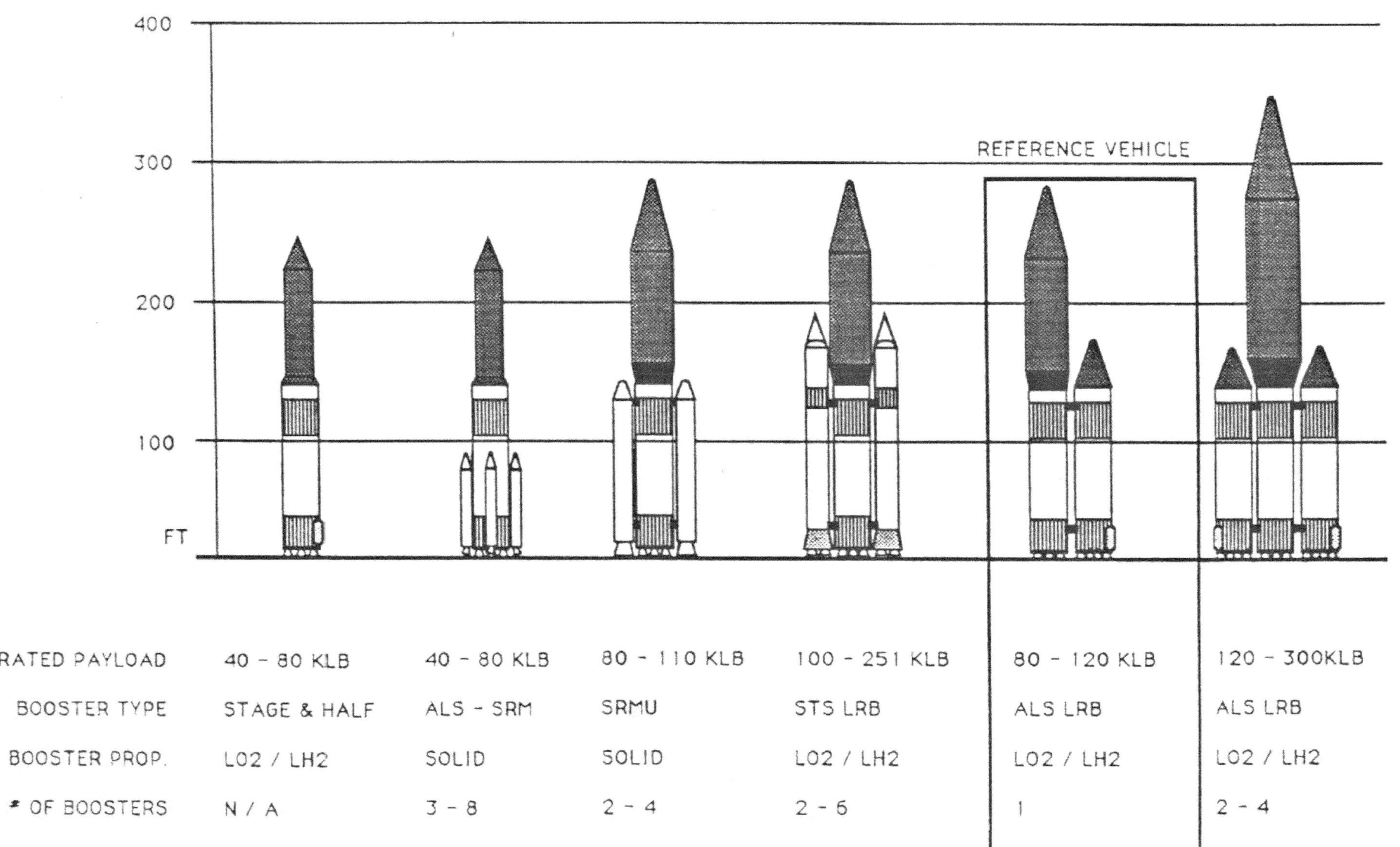

RATED PAYLOAD	40 - 80 KLB	40 - 80 KLB	80 - 110 KLB	100 - 251 KLB	80 - 120 KLB	120 - 300KLB
BOOSTER TYPE	STAGE & HALF	ALS - SRM	SRMU	STS LRB	ALS LRB	ALS LRB
BOOSTER PROP.	LO2 / LH2	SOLID	SOLID	LO2 / LH2	LO2 / LH2	LO2 / LH2
# OF BOOSTERS	N / A	3 - 8	2 - 4	2 - 6	1	2 - 4

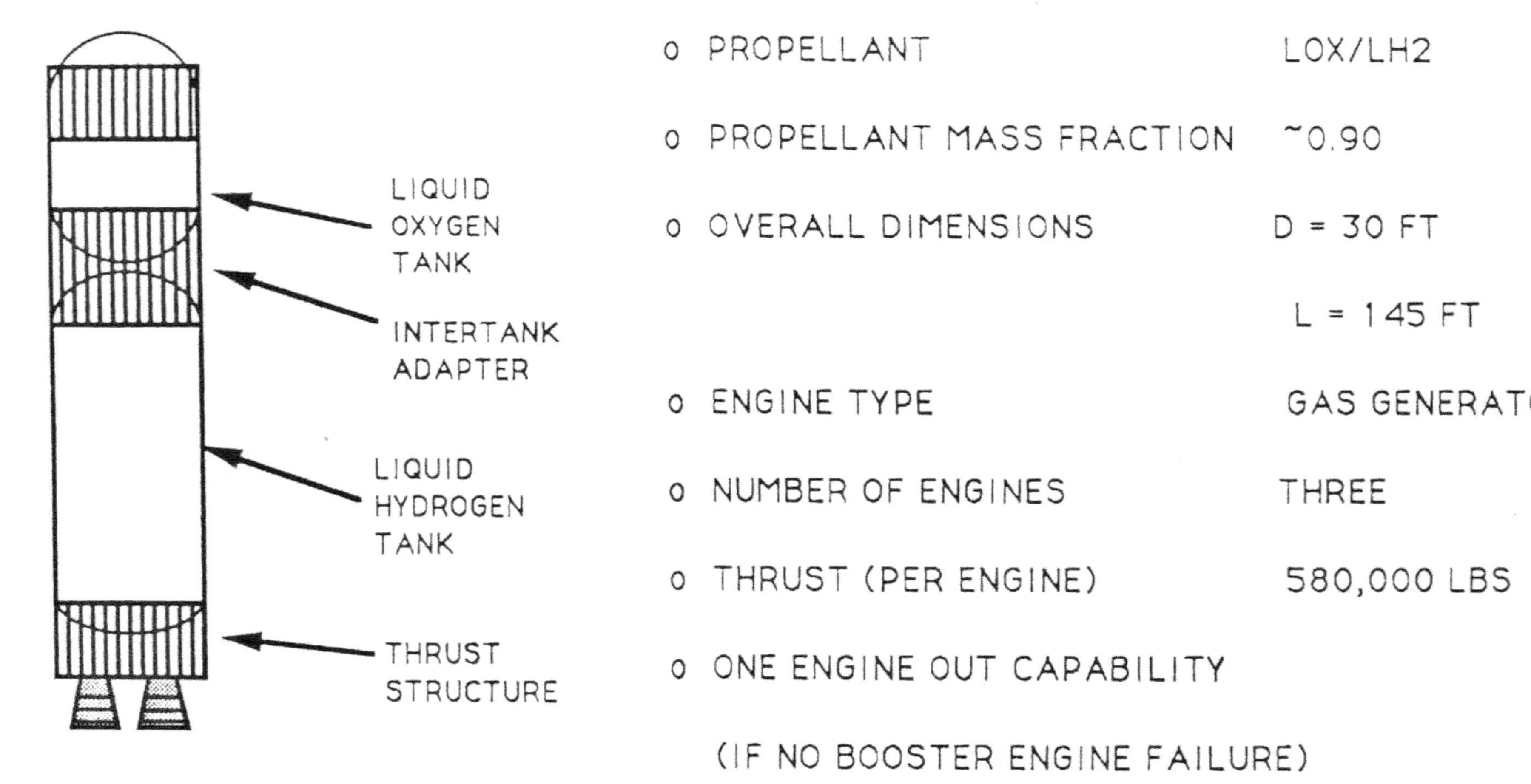
ALS COMMON CORE STAGE
LIQUID OXYGEN TANK
INTERTANK ADAPTER
LIQUID HYDROGEN TANK
THRUST STRUCTURE
3 CORE ENGINES
o PROPELLANT LOX/LH2
o PROPELLANT MASS FRACTION ~0.90
o OVERALL DIMENSIONS D = 30 FT
L = 145 FT
o ENGINE TYPE GAS GENERATOR
o NUMBER OF ENGINES THREE
o THRUST (PER ENGINE) 580,000 LBS
o ONE ENGINE OUT CAPABILITY
(IF NO BOOSTER ENGINE FAILURE)
o EXPENDABLE

ALS SOLID ROCKET MOTOR DESCRIPTION

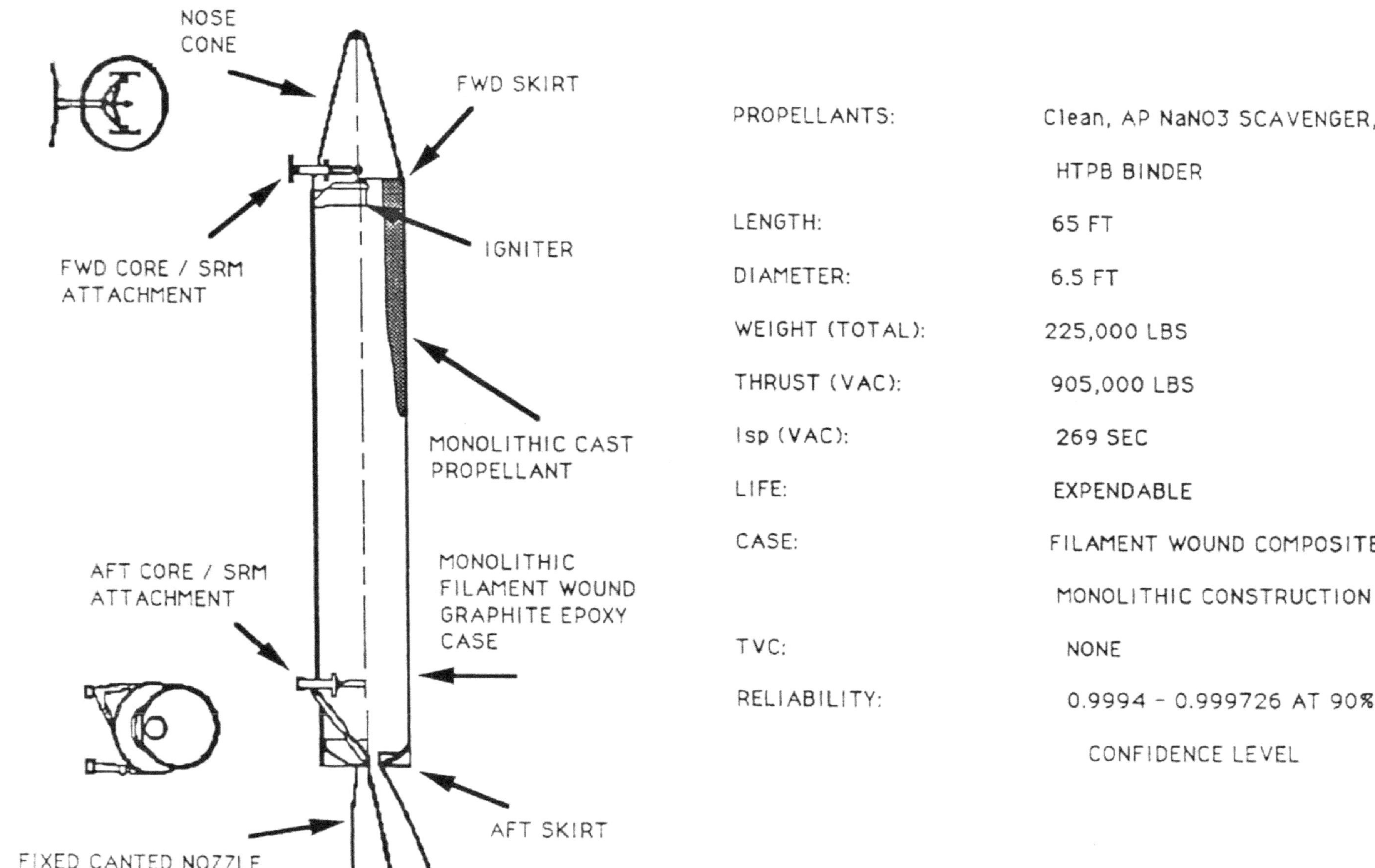

STRUCTURE AND FEED SYSTEM COMMONALITY

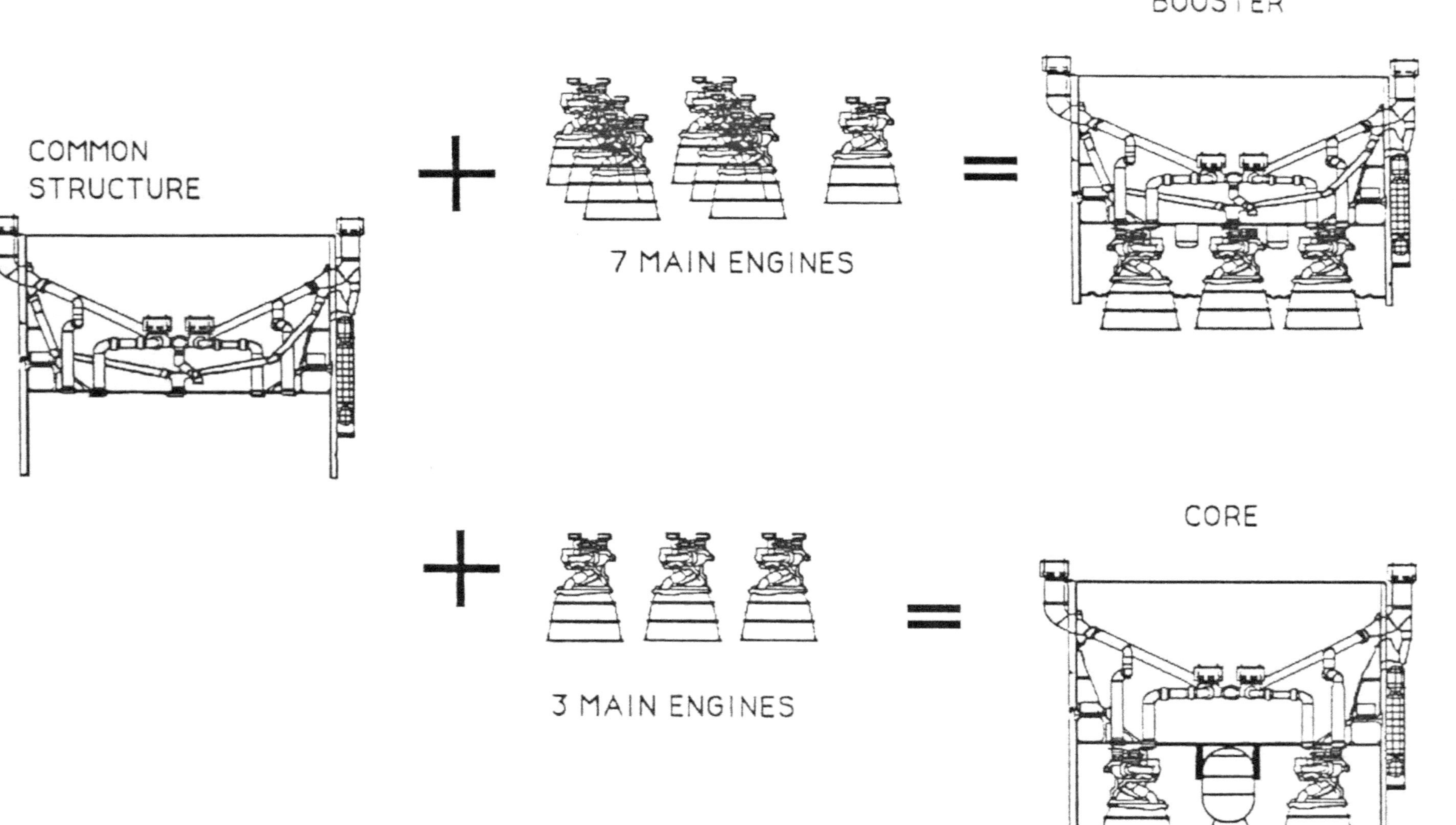

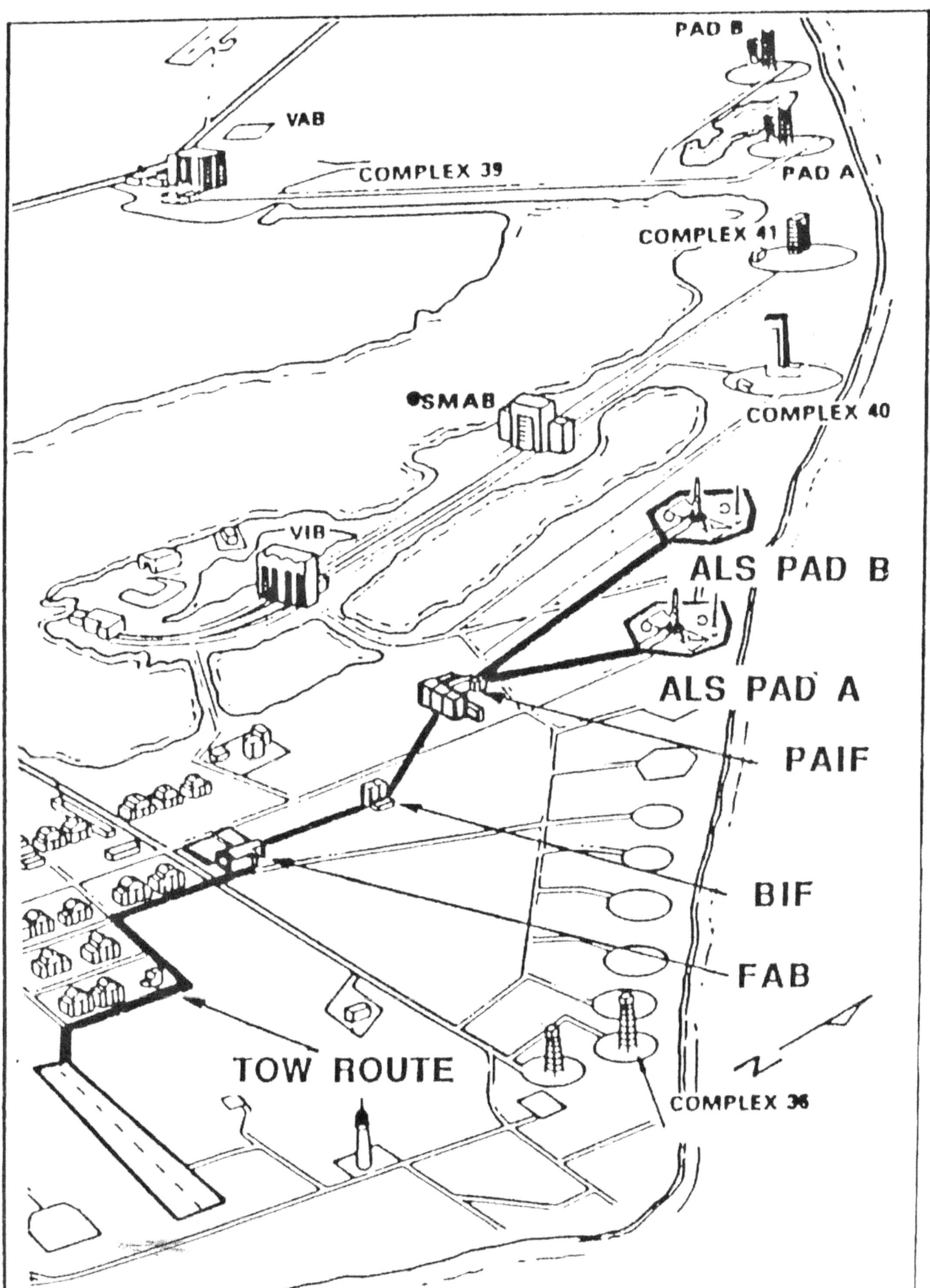

PAD B
VAB
COMPLEX 39
PAD A
COMPLEX 41
SMAB
COMPLEX 40
VIB
ALS PAD B
ALS PAD A
PAIF
BIF
FAB
TOW ROUTE
COMPLEX 36

www.ingramcontent.com/pod-product-compliance
Lightning Source LLC
Chambersburg PA
CBHW082108230426
43671CB00015B/2639
9781780398730